W0258979

Prof. Dr. Bernhard Thalheim

Born 1952 in Radebeul. From 1971 till 1989 at the Department of Mathematics of the University of Technology of Dresden, including study until 1975. Since 1989 full professor at Rostock University. Received Dr. rer. nat. in 1979 at Moscow State University, Dr. sc. nat. in 1985 at Dresden University.
Fields of interest: Theoretical Computer Science, Discrete Mathematics, Database and Knowledge Base Theory.

CIP-Titelaufnahme der Deutschen Bibliothek

Thalheim, Bernhard :
Dependencies in relational databases / Bernhard Thalheim. –
Stuttgart ; Leipzig : Teubner, 1991
(Teubner-Texte zur Mathematik ; Bd. 126)

NE: GT

ISBN 978-3-663-12019-3
DOI 10.1007/978-3-663-12018-6

ISBN 978-3-663-12018-6 (eBook)

TEUBNER-TEXTE zur Mathematik · Band 126
ISSN 0138-502X

Ursprünglich erschienen bei B. G. Teubner Verlagsgesellschaft mbH, Stuttgart · Leipzig 1991

Gesamtherstellung: Druckerei „G. W. Leibniz" GmbH, Gräfenhainichen

TEUBNER-TEXTE zur Mathematik · Band 126

Bernhard Thalheim

Dependencies in Relational Databases

Springer Fachmedien Wiesbaden GmbH 1991

An overview will be given in this book of the fundamental and the theoretical aspects of the foundations of the very promising relational database model. The book starts with an introduction to the notion of a relational database scheme, and to the logical and algebraic foundations of this relational model. The main topics of the book are the most important classes of constaints constituting an inherent property of database systems.
In this book different dependency classes are studied in detail: functional dependency, join dependency, inclusion dependency; generalizations of functional and join dependencies, hierarchical decomposition dependency, dependencies in relation with null values.
The book may also serve as a textbook for an advance course in relational database theory.

In diesem Buch wird ein Überblick über die fundamentalen und theoretischen Aspekte der Grundlagen des vielversprechenden relationalen Datenbankmodells gegeben. Nach der Einführung des relationalen Datenbankschemas werden die logischen und algebraischen Grundlagen des relationalen Datenbankmodells erläutert. Hauptgegenstand des Buches sind die bedeutendsten Klassen von Integritätsbedingungen, die die statischen Eigenschaften von Datenbanksystemen repräsentieren.
Detailliert werden im Buch verschiedene Abhängigkeitsklassen untersucht: funktionale Abhängigkeiten, Verbundabhängigkeiten, Inklusionsabhängigkeiten, Verallgemeinerungen funktionaler und Verbundabhängigkeiten, hierarchische Verbundabhängigkeiten und Abhängigkeiten in Relationen mit Nullwerten.
Das Buch kann als Einführung in die neueren Resultate der Theorie der relationalen Datenbanken genutzt werden.

Dans ce livre, on donne une vue d'ensemble des aspects fondamentaux et thêoriques des fondements du modêle de base de données relationnel qui s'avère três prometteur. Il dêbute par une introduction â la notion de schêma de base de données relationnelle et aux fondements algêbriques et logiques de ce modêle relationnel. Le principal sujet du livre est celui des classes de contraintes les plus importantes constituant une propueté inhérente des systêmes de base de données.
Dans ce livre, on étudie en detail differentes classes de dêpendances: les dêpendances fonctionnelles, les dêpendances relatives â l'operation de jonction, les dêpendances d'inclusion, les gêneralisations des dêpendances fonctionnelles ou relatives a l'operation de jonction, les dêpendances de composition hierarchique, les dêpendances relatives aux valeurs nulles.
On peut aussi utiliser ce livre comme manuel pour un cours de haut niveau sur la thêorie des bases données relationnelles.

В книге дан обзор о фундаментальных и теоретических аспектах основ многообещающей реляционной модели баз данных. Сначала вводится определение реляционной схемы баз данных и изучаются логические и алгебраические основы этой реляционной модели. Главным предметом книги являются самые важные классы семантических ограничений, которые образуют систему свойств свойственных для систем баз данных.
В этой книге детально изучены разные классы зависимостей: функциональные зависимости, соединительные зависимости, инклюзионные зависимости, обобщения функциональных и соединительных зависимостей, иерархические декомпозиционные зависимости и зависимости в отношениях с пустыми значениями.
Книга пригодна как введение в новейшие достижения теории реляционных баз данных.

PREFACE

> "It will be seen that logic can be used as a programming language, as a query language, to perform deductive searches, to maintain the integrity of data bases, to provide a formalism for handling negative information, to generalize concepts in knowledge representation, and to represent and manipulate data structures. Thus, logic provides a powerful tool for databases that is accomplished by no other approach developed to data. It provides a unifying mathematical theory for data bases."
>
> H. Gallaire, J. Minker April 1978

Today, database is a fascinating word. Commercial database management systems have been available for two decades, at the beginning in the form of hierarchical and network models. Two opposing research trends in database were created in the early seventies, the development of semantic database models and the introduction of the relational model. Most semantic data models were influenced by semantic networks. They are generally object-oriented and provide at least four types of primitive relationships between objects: classification (instance of), aggregation (part of), generalization (is-a), and association (member of). The relational model revolutionized the field by consequently separating data representation from underlying implementation what caused a reorientation in the methodology. Significantly, the inherent simplicity in the model permitted the development of powerful, non-procedural query languages and a lot of useful theoretical results. We confine our investigation to this model.

Generalized database management systems are considered as basic tools as programming languages, translators and operating systems. Nowadays much effort is devoted to establish a definite foundation of database technology in order to design more efficient and transparent systems and to enable optimization methods. By this understanding of the systems application will be improved as well. The philosophy behind database technology is sometime not quite understood because many users are not aware of the goals of database management systems. Consequently, these systems are often used wrong. The first step of the foundation of database theory is to be the precise definition of data models. Without a precise definition,a data model cannot be understood for purposes of the design, analysis, and implementation of schemata, transactions, and databases. A **database model** is a collection of mathematically sound concepts defining the intended structural and behavioral properties of objects involved in a database application. In the axiomatic approach, a database model is defined by the properties of its structures and operators. By the axiomatic approach conventional mathematics and logic were used to define the structural and behavioral properties of objects within the database model. Properties of data structures are given by axioms which are formal statements simple enough to be self-evident. Behavioral or dynamic properties are the operations that together with the data structures form the data model. Behavioral properties are given by inference rules which permit the deduction of the resultant properties for each meaningful database operation. In terms of logic, the semantics of each database within the database model can be deduced precisely by the application of valid inference rules to the set of axioms. Alternatively, the semantics of a syntactically correct schema are given by the axioms which characterize the databases to be accepted.

One of the most important database models is the relational model. One of the major advantages of the relational model is its uniformity. All data are seen as being stored in tables, with each row in the table having the same format. Each row in the table summarizes some object of relationship in the real world. The benefits and aims of the relational model are: to provide data schemes which are very simple and easily to be used; to improve logical and physical independence without references to the means of access to data; to provide users with high level languages which could be used by non-specialists in computing; to optimize access to the database; to improve integrity and confidentiality; to take into account a wide variety of applications; to provide a methodological approach for schema design and database design.

These benefits are based on a powerful theory the core of which is the theory of dependencies. Database dependencies can be regarded as a language for specifying the semantics of databases. They specify which of the databases are meaningful for the application and which of them are meaningless. Thus, the syntactic specification is joined with semantic specification. Dependencies constitute an inherent property of database systems. They express the different ways by that data are associated with one another. Since many different associations of data exist, a lot of different classes of dependencies (more than 90) are considered in more than thousand papers. For some classes the implication problem is solved. By studying their respective properties it can be shown how different types of dependencies interact with one another. These properties may be considered as inference rules which allow to deduce new dependencies as well as to generate the closure of all dependencies. Solving this problem, we can test whether two given sets of dependencies are equivalent or whether a given set of dependencies is redundant. A solution for these problems seems to be a significant step towards automated database schema design, towards automated solution of the above-mentioned seven aims and towards recognizing computational feasible problems and the unfeasible ones.

At present we know at least five fields of application of dependency theory:

(1) normalization for a more efficient storage, search and modification;

(2) reduction of relations to subsets with the same information together with the semantic constraints;

(3) utilization of dependencies for deriving new relations from basic relations in the view concept or in so-called deductive databases;

(4) verification of dependencies for a more powerful and user-friendly, nearly natural language design of databases;

(5) transformation of queries into more efficient search strategies.

Other important applicabilities of the relational database theory are in other branches of computer science, in discrete mathematics, in most of other database models, in optimization, in pattern recognition and in algebra. Because we want to present an unifying approach to dependency theory and intend only to give an orientation for literature, some branches of relational database theory as the theory of relational algorithms, theoretical foundations of query languages, optimization and normalization are only briefly cited.

This book comprises 9 sections. In section 1, the basic database terminology is presented. Section 2 describes elementary database operations. A theoretical discussion of dependency theory is given in section 3 where emphasis is laid the various logical problems of database theory. Sections 4, 5, 6 deal with the most important classes of dependencies, the propositional dependencies, a subclass of which is the class of functional dependencies, join dependencies and inclusion dependencies. In section 7, several existing approaches to dependency theory for relations with null values are described and compared. Other dependencies used for horizontal decomposition of relations are discussed in section 8. Finally, several topics designated for future research are described in section 9.

I would like to thank the Teubner Publishing House for the publication of this monograph. In addition thanks should be expressed to the collegues in Dresden, Berlin, Moscow and Budapest for useful discussions and to Mrs. Scheller for the grammatical inspection of the manuscript. Above all, I wanted to thank my wife, Valeria, for their assistance, support and understanding.

Dresden, December 1986, Kuwait, 1988 Bernhard Thalheim

C O N T E N T S

1. Database Schemes and Databases 7

1.1. The Relation Scheme and Relational Databases 7
1.2. The Entity-Relationship Model 16

2. The Relational Algebra 25

2.1. The Algebraic Language 25
2.2. Relational Expressions 31
2.3. Algebraic Dependencies 33

3. Some Fundamentals of Dependency Theory 35

3.1. Logical fundamentals of Dependency Theory 35
3.2. Dependencies 41
3.2.1. Logical Dependencies 44
3.2.2. Special Algebraic Dependencies 47
3.2.3. A Proof Procedure for General Implicational Dependencies 49
3.3. Template Dependencies and Tuple-Generating Dependencies 51
3.4. Embedded Dependencies 55
3.5. General Functional Dependencies 60
3.6. The Deductive Basis of Relations 63
3.7. Design By Example 68

4. Functional Dependencies 72

4.1. Properties of Generalized Functional Dependencies 73
4.2. Properties of Functional Dependencies 86
4.3. Hungarian and Monotone Functional Dependencies 97
4.4. Key Dependencies 102
4.5. Armstrong Databases 114
4.6. Degenerated Multivalued Dependencies 121

5. Join Dependencies 125

5.1. Multivalued Dependencies and Binary Join Dependencies 127
5.2. Full Hierarchical Dependencies and Acyclic Join Dependencies 138
5.3. The Class of Join Dependencies 144

6. Inclusion Dependencies 153

6.1. The Class of Inclusion Dependencies 153
6.2. Inclusion Dependencies and Their Interaction with Functional Dependencies 159

7. Dependencies in Relations with Null Values and Incomplete Informations 167

7.1. Databases with Null Values 170
7.2. Databases with Incomplete Information 176
7.3. Context-Dependent Null Values 179
7.4. Key Sets in Relations with Null Values 181

8. Horizontal Decomposition Dependencies 187

8.1. The Horizontal Decomposition 187
8.2. Conditional Functional Dependencies 190
8.3. Union Constraints 194

9. The Relationship between Dependency Classes 197

References 202

Index 216

1. DATABASE SCHEMES AND DATABASES

1.1. THE RELATION SCHEME AND RELATIONAL DATABASES

We attempt a more rigorous definition of the relational database model based on /THAL 88/ as it was originally introduced by E.F. Codd /CODD 70/ using the theory of abstract data types /REI 84/ and especially the approach of /PDGG 88/, /VOSS 87/ and /DEAB 85/. The underlying concept used in the relational model is the same as that used to define a mathematical relation (in set theory and algebra). Simply, a relation is a subset of the Cartesian products of a list of domains, a domain being merely a set of entity values.

From the algebraic point of view, a relation can also be understood as a set of functions from domain names in domains. This point of view allows short and clear definitions. We will also compare these approaches and use one of them in different chapters.

In the relational model, it is essential to make a distinction between two different levels: the intention or meaning of a relation and the extension or realization of a relation as a set of tuples (or functions) which comes up to the rules by its intention. Using the relational vocabulary, the words relation and relational database are used to designate an extension, and the words relational scheme and database scheme to designate its corresponding intention.

A <u>relational database scheme</u> RS = (U , $\underline{D}$, dom) (or shortly **relation scheme**) is given
by a finite set U of so-called **attributes** (or sort names (universal algebra approach) or column names (representation of relations by tables)),
by a set $\underline{D}$ = $\{D_1, D_2, \ldots\}$ of **domains**,
and by an arity or **domain function** dom : U ⟶ $\underline{D}$ which associates with every attribute its domain.
Note that in difference to the classical approach we use a strongly many-sorted approach which claims that the same attribute can not be used twice for columns in tables.
It is useful to utilize a shorter notion for relation schemes. If $\underline{D}$ and dom are obvious or defined by the context or arbitrary ($\underline{D}$={set_of_all_strings}) or not of importance for the topic under consideration then $\underline{D}$ and dom are omitted.

A **tuple** on RS = $(U,\underline{D},dom)$ is a function $t : U \longrightarrow \bigcup_{D \in \underline{D}} D$ with $t(A) \in dom(A)$ for $A \in U$. If there is defined an order on U ($U = \{A_1,A_2,\ldots A_n\}$ then the tuple can be represented by $(t(A_1),\ldots,t(A_n))$.
We denote by T(RS) the set of all tuples on RS.

Any subset r of T(RS) is called **relation** (on RS).

A given sequence DRS = $RS_1,RS_2,\ldots RS_m$ of relation schemes is called **compatible** if it holds the property $dom_i(A) = dom_j(A)$ for $A \in U_i \cap U_j$ where $RS_i = (U_i,\underline{D}_i,dom_i)$.

For a compatible sequence of relation schemes there can be defined a common function dom with $dom_i(A) = dom(A)$ for $A \in U_i$.

For a given compatible sequence DRS = $RS_1,RS_2,\ldots RS_m$ of relation schemes and a function $C : Pow(T(RS_1)x\ldots x\ T(RS_m)) \longrightarrow \{0,1\}$
a **database scheme** DS is the pair (DRS , C)
where by Pow(M) is denoted the power set of M.
The function C is called **integrity constraint**.
For a given database scheme DS = ($RS_1,\ldots,RS_m$, C) a **DS-relational database** (or shortly **DS-database** or **database** if DS is defined by the context) is given by the family $(r_1,\ldots r_m)$ where the r_i are relations on RS_i $(1 \leq i \leq m)$ and $C(r_1,\ldots,r_m) = 1$.

Let us now consider some examples.
Example 1.1. Suppose we are intended to handle some informations about our friends. We are interested in their first and their last name, the address, the telephone number and their main hobby. This information can be stored in a relation FRIENDS which contains six columns headed by NAME, FIRST_NAME, TOWN, STREET, PHONE_NUMBER, HOBBY. All the columns contain strings. Therefore we can define:
U = {NAME, FIRST_NAME, TOWN, STREET, PHONE_NUMBER, HOBBY},
$\underline{D}$ = {set of all strings},
the function dom associates the set U with the set of all strings.
The function C contains at least the condition that if the addresses are different for two friends then the phone numbers are also different.
Then we define the database scheme FRIENDS = $((U,\underline{D},dom),C)$.

Example 1.2. Now we give a not so small example of a database scheme. Consider now the hotel database of /PDGG88/ which contains different information on the rooms in the hotel, the employees, the visitors, the stays and the phone-bills. Therefore let

U_1 = {ROOM-NUMBER, BEDs-NUMBER, FLOOR, RATE, TV?, BATH?};

$\underline{D}_1$ = {set of room numbers, set of positive integers, {true,false}},

dom_1 is straightforward. The set of positive integers is associated with BEDs-NUMBER, FLOOR, and RATE. The set of truth values is associated with the two questions on tv and bath room for the hotel room;

ROOMS = $(U_1,\underline{D}_1,dom_1)$;

U_2 = {EMPLOYEE-NUMBER, E-NAME, JOB, SALARY};

$\underline{U}_2$ and dom_2 are obvious;

EMPLOYEES = $(U_2,\underline{D}_2,dom_2)$;

U_3 = {VIS-NUMBER, VIS-NAME, VIS-STREET, VIS-CITY, VIS-COUNTRY};

$\underline{U}_3$ and dom_3 are obvious;

VISITORS = $(U_3,\underline{D}_3,dom_3)$;

U_4 = {VIS-NUMBER, ARRIV-DATE, LEAV-DATE, ROOM-STAY, BILL};

$\underline{U}_4$ and dom_4 are obvious;

STAYS = $(U_4,\underline{D}_4,dom_4)$;

U_5 = {ROOM-NB, TIME, DATE, DESTINATION, PHBILL, PAID?};

$\underline{U}_5$ and dom_5 are obvious;

PHONE-BILLS = $(U_5,\underline{D}_5,dom_5)$;

C can include different conditions such as:

- every room has a different number,
- there are only 5 floors and the first digit of the room number indicates the floor,
- every room in floor 1 has a bath,
- all employees have different numbers,
- every visitor have a different number,
- if two visitors live in the same town, then the country is the same,
- a visitor leaves on a later date than his arrival date,
- a visitor cannot phone at the same time twice,
- the rooms where visitors stay are rooms of the hotel,
- the rooms of the phone bills are rooms of the hotel,
- if there is a phone call from a room then that room was occupied that date.

Now let HOTEL be the following database scheme

(ROOMS, EMPLOYEES, VISITORS, STAYS, PHONE-BILLS, C) .

The function C is defined here in an abstract way. But for our purposes, this function can be defined using a logical language.

Given a compatible sequence DRS = $RS_1,\dots,RS_m$ of relation schemes with RS_i = $(U_i,\underline{D}_i,dom_i)$ and $\underline{D}_i = \{D_{i1},\dots,D_{il}\}$ $(1 \le i \le m)$.

Then we use the following alphabet ALPH(DRS) :

VAR(A) - set of all variables for the attribute A

CONST(A) = { c' ¦ c ∈ dom(A) } - set of all constants for the attribute A

VARCONST(A) = VAR(A) + CONST(A)

$P_1,\dots,P_m$ - corresponding predicates for the relation schemes

¬ (negation), ∧ (conjunction), v (disjunction), ==> (implication), <==> (equivalence), ∀ (generalization), ∃ (particularization), parentheses, comma.

Let VAR be the set of all variables. For our purposes, we assume that this set is unique for all alphabets and that this set is covered by the sets VAR(A).

A term is a variable or a constant.

The string x = y for x ∈ VAR(A), y ∈ VARCONST(B) with dom(A)=dom(B) is called equality formula.

For $U_i = \{A_1,\dots,A_n\}$ the string $P_i(x_1,\dots,x_n)$ with x_i ∈ VARCONST(A_i) is called predicate formula.

The set L(DRS) of **formulas** on DRS is defined as follows:

1. Equality formulas and predicate formulas are formulas.
2. If F and F' are formulas, and x is a variable, then (¬F), (F∧F'), (F v F'), (F ==> F'), (F <==> F'), ∀x F , ∃x F are formulas.
3. An expression is a formulas if it can be shown to be a formula on the basis of clauses 1. and 2.

We use the usual conventions to omission of parentheses that ∀, ∃, <==>, ==>, ¬, ∧, v rank in strength in this order.

Using these definitions, we can introduce inductively the set of **free variables** of formulas from L(DRS).

1. For F = $P(x_1,\dots,x_n)$ ∈ L(DRS) let Fr(F) be the set $\{x_1,\dots,x_n\}$.
2. For F = x=y , F' = x=c let Fr(F)={x,y} , Fr(F')={x} .
3. For F = (¬F') Fr(F) = Fr(F') .
4. For F = (F' * F") and * ∈ { ∧, v, ==>, <==>} Fr(F) = Fr(F') ∪ Fr(F").
5. For F = QxF' , Q∈ {∀,∃} , Fr(F) = Fr(F') - {x} .

It is possible to use a more understandable notion in formulas. For instance, $P(x_1,\dots,x_n)$ can be denoted by $P(\hat{x})$ or $P(\hat{y},\hat{z})$ for sequences of variables $\hat{x} = x_1,\dots,x_n$, $\hat{y} = y_1,\dots,y_m$, $\hat{z} = z_1,\dots,z_k$ with $\{y_1,\dots,y_m\}$

$\{z_1,\ldots,z_k\} = \{x_1,\ldots,x_n\}$ (It is not excluded, that $\{y_1,\ldots,y_m\} \cap \{z_1,\ldots,z_k\} \neq \emptyset$.). The notion $\hat{x}=\hat{y}$ means the formula $x_1=y_1 \wedge x_2=y_2 \wedge \ldots \wedge x_m=y_m$ for $\hat{x}=x_1,\ldots,x_m$ and $\hat{y}=y_1,\ldots,y_m$. A formula $F = \forall x_1 \forall x_2 \ldots \forall x_m F'$ where F' is quantifier-free and $Fr(F')=\{x_1,\ldots,x_m\}$ is called **universal formula** and denoted shortly by $\forall(F')$. For sequences of variables $\hat{x}=x_1,\ldots,x_m$, $\hat{y}=y_1,\ldots,y_k$ a formula $\forall x_1\ldots\forall x_m \exists y_1\ldots\exists y_k(F)$ will be denoted by $\forall\hat{x}\exists\hat{y}(F)$.
If there is impossible a misunderstanding or confusion we write x instead of $\hat{x}$.

Using these definitions, the notion of a database scheme can be introduced more concrete. For a given compatible sequence DRS = $RS_1,RS_2,\ldots RS_m$ of relation schemes and a set of formulas Form from L(DRS), a **database scheme** DS is the pair (DRS,Form). The set Form is also called **integrity constraints**. Only such databases are considered for DS in which the integrity constraints from Form are valid, i.e. for a given database scheme DS = $(RS_1,\ldots,RS_m$,Form) a **DS-database** by the family $(r_1,\ldots r_m)$ where the r_i are relations on RS_i $(1\leq i\leq m)$ and the formulas from Form are valid.
By R(DS) we denote the class of all DS-databases.

Now we define the validity of formulas.

In semantics we are concerned with interpretations where an interpretation of a set of formulas includes the specification of a non-empty set (or domain) D from which variables are given values. For databases, the set D is predefined by the scheme.
Let DRS = $RS_1,RS_2,\ldots,RS_m$ be a sequence of compatible relation schemes (RS_i = $(U_i,\underline{D}_i,dom_i)$, $U = \bigcup_{i=1}^{m} U_i$, dom the domain function of DRS, and $D = \bigcup_{A\in U} dom(A)$).
Let further $M=(r_1,\ldots,r_m) \in Pow(T(RS_1)\times\ldots\times T(RS_m))$.
Any mapping I : VAR ⟶ D which is compatible with the attribute separation, i.e. $I(x) \in dom(A)$ for $x \in VAR(A)$, is called **interpretation** for the variables in D .
We can extend the interpretation in an obvious way to DRS-formulas. Let I:VAR⟶D be an interpretation for VAR. We define recursively, what does it mean when M **satisfies** $F \in L(DRS)$ under the interpretation I (i.e. that F is satisfied in M for I, denoted by $M\models F[I]$):

a) If $F = P_i(x_1,\ldots,x_n)$ then $M\models F[I]$ iff $(I(x_1),\ldots,I(x_n)) \in r_i$.
b) If $F = x=c'$, then $M\models F[I]$ iff $I(x) = c$.
c) If $F = x=y$, then $M\models F[I]$ iff $I(x) = I(y)$.
d) If $F = \neg F'$, then $M\models F[I]$ iff it is not true that $M\models F'[I]$.

e) If $F = F' \wedge F''$, then $M||==F[I]$ iff $M||==F'[I]$ and $M||==F''[I]$.

f) If $F = F' \vee F''$, then $M||==F[I]$ iff $M||==F'[I]$ or $M||==F''[I]$.

g) If $F = (F' ==> F'')$, then $M||==F[I]$ iff $M||==F''[I]$ or it is not true that $M||==F'[I]$.

h) If $F = (F' <==> F'')$, then $M||==F[I]$ iff $M||==F'[I]$ if and only if $M||==F''[I]$.

i) If $F = \forall x F'$, then $M||==F[I]$ iff for every interpretation I' of VAR which differs from I only on x one has $M||==F'[I']$.

j) If $F = \exists x F'$, then $M||==F[I]$ iff for some interpretation I' of VAR which is different from I only on x $M||==F'[I']$.

A DRS-formula F is said to be **valid in M** (i.e. that M is a **model** of F, denoted by $M||==F$) if $M||==F[I]$ for every interpretation I:VAR—>D . A set of DRS-formulas Form is said to **valid** in M (i.e. that M is a **model** of Form, denoted by $M||==Form$) if it holds $M||==F$ for any $F \in Form$.

A DRS-formula F **follows** from a set of DRS-formulas Form , denoted by Form $|=$ F if F is valid in all models of Form .

If a relation or a database is the realization of a scheme the notion of relation or database corresponds to a certain situation in the database. The set R(DS) is therefore the set of possible states of the relational database scheme DS . Consequently, a dynamical database can be defined as a sequence $M_1, M_2, \ldots, M_i, \ldots$ of DS-databases for some relational database scheme DS .

Usually, if there cannot be a misinterpretation, we apply the notion $r||==F$ or $(r_1,\ldots,r_m) ||== F$ instead of $M ||== F$.

Example 1.3. Consider the following description of a Cinema information concerning the following entity sets:

- C (inema) - A (ddress) - T (ime)
- F (ilm) - P (roducer) - M (ain actor) .

We get the relation scheme RS = (U,{Set of all strings},dom) with U = {C,A,T,F,P,M}. Now the set of DRS-formulas Form = $\{F_1,F_2,F_3\}$ and a DRS-formula F_4 are given:

$F_1 = P(c,a',t',f',p',m') \wedge P(c,a,t,f,p,m) \longrightarrow a = a'$;

$F_2 = P(c,a',t,f',p',m') \wedge P(c,a,t,f,p,m) \longrightarrow f = f'$;

$F_3 = P(c',a',t',f,p',m') \wedge P(c,a,t,f,p,m) \longrightarrow p = p'$;

$F_4 = P(c,a',t',f,p',m') \wedge P(c,a,t,f,p,m) \longrightarrow a = a' \wedge p = p'$.

Obviously, we get Form $|=$ F_4 .

Example 1.4. In this text, we have been using a part of an university management system. The database includes a table of courses with the attributes and lecturer, a timetable with the attributes of lecture, term, time, room, a table of students with the attributes of student's name, address and term and a table of marks with the attributes of lecture, student's name, year the mark was given and mark.

Now we establish RS_1 = COURSE = (U_1,$\underline{D}$,dom_1)

RS_2 = TIMETABLE = (U_2,$\underline{D}$,dom_2)

RS_3 = STUDENT = (U_3,$\underline{D}$,dom_3)

RS_4 = MARKS = (U_4,$\underline{D}$,dom_4) where

$\underline{D}$ = {set of all strings} ,

U_1 = {LECTURE, LECTURER} ,

U_2 = {LECTURE, TERM, TIME, ROOM} ,

U_3 = {NAME, ADDRESS, TERM} ,

U_4 = {LECTURE, NAME, YEAR, MARK} ,

and dom_1, dom_2, dom_3, dom_4 are obvious.

The set Form with

$\forall x,y,z,u \; \exists v$ (timetable(x,y,z,u) —> course(x,v)) ,

.(timetable(x,y,z,u) ∧ timetable(x',y',z,u) —> x,y = x',y') ,

.(student(w,v,u) ∧ student(w,v',u') —> v,u = v',u') is given.

Let now DS = UNIVERSITY = (RS_1,RS_2,RS_3,RS_4, Form).

The following database is a UNIVERSITY-database.

LECTURE	LECTURER
computer science	Bachmann
algebra/geometry	Bormann
logic	Thiele
analysis	Mulla
databases	Thalheim

LECTURE	TERM	TIME	ROOM
computer science	1	tu 1	Kh4 123
algebra/geometry	1	sa 2	Ad1 234
analysis	3	mo 1	Kh1 345
logic	7	we 3	Kh7 456
databases	9	we 2	Ja1 567

NAME	ADDRESS	TERM
Schulze	Dresden	1
Farouk	Kuwait	3
Hani	Detroit	5
Ruslan	Sofia	7

LECTURE	NAME	YEAR	MARK
analysis	Schulze	1986	A
analysis	Farouk	1985	B
algebra/geometry	Ruslan	1986	D
algebra/geometry	Hani	1988	F.

We can define for a DS-database also its logical theory.
Let $DS = (RS_1,\ldots,RS_m,Form)$ a database scheme where $RS_i = (U_i,\underline{D}_i,dom_i)$,
$U = \bigcup_{i=1}^{m} U_i$, $\underline{D}_i = \{D_{i1},\ldots,D_{i\,l(i)}\}$, $D = \bigcup_{A \in U} dom(A)$.
We define now for a given tuple $M = (r_1,\ldots,r_m)$ of relations on DRS
$DIS_{DS} = \{- c' = d' \mid c,d \in D, \ c \neq d\}$,
$Form_{M,i} = \{P_i(c'_1,\ldots,c'_m) \mid (c_1,\ldots,c_m) \in r_i\}$
$\cup \{-P_i(c'_1,\ldots,c'_m) \mid (c_1,\ldots,c_m) \notin r_i\}$ $(1 \leq i \leq m)$,
$Form_M = \bigcup_{i=1}^{m} Form_{M,i}$.
The set $DIS_{DS,M}$ $Form_M$ is called the **diagram** of M .

<u>Corollary 1.1.</u> For any set of DRS-formulas Form M $\models\!=$ Form iff DIS_{DS}+$Form_M$+Form is satisfiable.

Using these definitions, it is also possible to introduce the concepts of inclusion and equivalence between schemata.
Intuitively, two schemata DS = (DRS,Form) , DS' = (DRS',Form') are equivalent if for each DS-database M a DS'-database M' exists from which we can extract exactly the same information and vice versa. This concept can be understood as the concept of behavioral equivalence and may be formalized saying that for each query q on M a query q' on M' must exist such that they give exactly the same answer. In /AUBM 80/ it has been shown that this condition holds if and only if a query on M exists whose result is M' and a query on M' exists whose result is M . Our definitions are based on this last property. Regarding the inclusion of schemes, we may be interested in two kinds of situations:
- for each DS-database M a DS'-database M' exists that contains at least the same information;
- for each DS-database M a DS'-database M' exists that contains exactly the same information.

These two situations arise, respectively, when we wanted to know whether a decomposed scheme looses any information. As a consequence, we give two definitions of inclusion between schemes.

Given a database scheme DS = (DRS,C) , $DRS = RS_1,\ldots,RS_k$, and sets of DRS-formulas. Given further a DS-database $M = (r_1,\ldots,r_k)$.
Now we can define the "value" of formulas according to M : Given a DRS-formula F with $Fr(F) = \{x_1,\ldots,x_m\}$. Then

$F(M) = \{ (t_1,\ldots,t_m) \mid$ for some interpretation I $M \models F[I]$ and
$t_j = I(x_j)$, $1 \leq j \leq m$ $\}$.

Given two database schemes $DS = (DRS,C)$, $DRS' = (DRS',C')$, $DRS = RS_1,\ldots,RS_k$, $DRS' = RS'_1,\ldots,RS'_l$, sets of DRS-formulas and of DRS'-formulas.

(1) DS is **weakly included** in DS' (denoted by $DS \leq DS'$) (with respect to the sets of formulas) if DRS-formulas $F_1,\ldots,F_l$ exist such that for any DS-database M a DS'-database $M' = (r'_1,\ldots,r'_l)$ exists such that for any i, $1 \leq i \leq l$, $r_i = F_i(M)$.

(2) DS is **included** in DS' (denoted by $DS -\!< DS'$) (with respect to the given formulas) if there exist DRS-formulas $F_1,\ldots,F_l$ and DRS'-formulas $F'_1,\ldots,F'_k$ such that for any DS-database $M=(r_1,\ldots,r_k)$ a DS'-database $M'=(r'_1,\ldots,r'_l)$ exists such that for any i,j, $1 \leq i \leq l$, $1 \leq j \leq k$, $r'_i = F_i(M)$ and $r_j = F'_j(M')$.

(3) DS is **weakly equivalent** to DS' if $DS \leq DS'$ and $DS' \leq DS$.

(4) DS is **equivalent** to DS' if $DS -\!< DS'$ and $DS' -\!< DS$.

In the case of scheme inclusion $((F_1,\ldots,F_l),(F'_1,\ldots,F'_k))$ is called lossless scheme transformation.

There are many lossless scheme transformations, among which two algebraic transformations (projection/join (chapter 5), selection/union (chapter 8)) and one logical transformation (reduction/cover (chapter 3.4)) are dealt with in this book. Views /DEAB 85/ are clearly modeled by weak inclusion. Lossless vertical decomposition is modeled by inclusion but, in general, not by equivalence. Dependency preserving vertical decomposition is modeled by inclusion. Lossless vertical decomposition with hidden dependencies /SMSM 77/ is modeled by equivalence. Hierarchical decompositions are modeled by equivalence.

<u>Example1.5.</u>Let $DS = ((\{1,2,3\}),C)$ and $DS' = ((\{1,2\}),(\{1,3\}),C')$.
If C is composed of a formula $.(P(x,y,z') \wedge P(x,y',z) \Longrightarrow P(x,y,z))$ and C' is composed of two formulas
$\forall x \forall y \exists z (Q_1(x,y) \Longrightarrow Q_2(x,z))$ and $\forall x \forall z \exists y (Q_2(x,z) \Longrightarrow Q_1(x,y))$ then the pair of transformations $((\exists z P(x,y,z),\ \exists y P(x,,z)),\ (Q_1(x,y) \wedge Q_2(x,z))$ becomes lossless. The schemes DS and DS' are equivalent.
If DS, DS', C' are the above and $C = \emptyset$ then we get $DS \leq DS'$ using the transformation $(\exists z P(x,y,z),\ \exists y P(x,y,z))$.
If C is composed of two formulas
$.(P(x,y,z) \wedge P(x,y',z') \longrightarrow y = y')$ and $.(P(x,y,z) \wedge P(x',y,z') \longrightarrow z = z')$ and

C' is composed of two formulas

$.(Q_1(x,y)\text{-}Q_1(x,y') \longrightarrow y=y')$ and $.(Q_2(x,z)\text{-}Q_2(x,z') \longrightarrow z=z')$

we obtain $DS = ((\{1,2,3\}),C) \prec DS' = ((\{1,2\}),(\{1,3\}),C')$.

In this example U = {EMPLOYER,CITY,ZIP} can be understood as a concretization of U = {1,2,3} .

1.2. THE ENTITY-RELATIONSHIP MODEL

The classical Relational Model deals only with flat relations. It is not aware of any distinction between entity relations and relationship relations. In contrast, models like the network model and the hierarchical model make distinctions between these two types of relations. In practical database design, such distinctions can often be perceived intuitively.

The Entity-Relationship Model (ERM) has been recognized as an excellent tool for high level database design because of its many convenient facilities for the conceptual modeling of reality. Its basic version /CHEN76/ deals with more static properties, such as entities, attributes and relationships. More recently considerable effort has been devoted to query manipulation capabilities, to theories modeling more semantic knowledge and to related theories. These attempts arise from practical needs and from the common feeling that the relational model facilities can be and should be generalized for more complex data models. One of the main objectives of the relational model is communicability, which means offering the user a data model which is easy to understand, use and communicate about. Regretfully, this objective is only partially fulfilled by the relational model since it conceals much of the semantic structure of the real world. ERM reflects a natural, although limited, view of the world: entities are qualified by their attributes and interactions between entities are expressed by relationships. Codd pointed out /CODD 82/ that the semantic data models in general, and ERM in particular, lack both a well defined instance level and, therefore, a well defined data manipulation language. The ERM has been mostly accepted as an early stage data base design tool. Once the design stage ends, the entity-relationship scheme, represented by an entity-relationship diagram is translated into a relational scheme, or a network scheme and its role is therewith ended /ULLM82/. We don't agree completely with this point of view. The semantic information enclosed in the

ERM should be used further, especially for normalization and query optimization. By contrast, the theoretical assumptions of the relational model are commonly accepted. This is expressed in Chen's proposal of developing a special algebra for ERM /CHEN84/, as well in /SUMI87/. Indeed, majority of the database community still believes that the relational model paradigms (in particular, the relational algebra (chapter 2) and logic (chapter 3)) are successful as an intellectual tool for the database domain. Thus there is a great temptation to extend this success to other database ideas that are badly in want of a solid theoretical basis. Examples of this effort are "database logic" /JACO82/ which may be applied to hierarchical and network models /DEAB85/, and "multimodel database systems" /MAPI82/, another calculus-oriented approach to specification of query languages for richer models. There are two obstacles for such extensions of the relational theory. First, ERM has plenty of persistent concepts (such as relationships with attributes, multivalued attributes, attributes having subattributes, duplicates, ordering, "is-a" generalizations, and so on) which are very hard to formalize within theory of relations or within formal logic. Second, the relational algebra and the logic are inconsistent with respect to specification of query languages. Duplicates which can be returned by a query in current languages like SQL and QUEL, ordering, updating operations, and a lot of other operators (aggregate, arithmetic, transitive closure) are not covered by the relational algebra and are not expressible in a homogeneous way in pure relational calculi.

The database literature introduces many definitions of the concept of data model. Codd /CODD81/ advocates a kind of equivalence between data models and data structures (together with operations and constraints). Brodie views as /THAL84/ a data model as a collection of mathematically well defined concepts. The ERM was originally designed to be a description of a very informal world for people who want to understand it, thus this scheme does not necessarily have to be formalized, and it really describes the world and not data structures. But it is impossible to define the mapping of an ERM to another model without formalization of data structures which are to be queried and manipulated in the new model. Therefore we introduce in a formal approach the entity-relationship scheme and the entity-relationship diagram.

A **data scheme** DD = (U , $\underline{D}$, dom) is given
by a finite set U of **attributes** ,
by a set $\underline{D} = \{D_1, D_2, \ldots\}$ of **domains**,
and by an arity or **domain function** dom : U $\longrightarrow$ $\underline{D}$ which associates with every at-

tribute its domain.
Note that in difference to the classical approach we use a scheme of data first and then we define the corresponding schemes.

A **tuple** on $X \subseteq U$ and on DD = (U,$\underline{D}$,dom) is a function $t : X \longrightarrow \bigcup_{D \in \underline{D}} D$ with $t(A) \in dom(A)$ for $A \in X$.
Given now a set of tuples r on X and DD , and a subset Y of X . Y is called **key** of r if all elements of r can be distinguished using Y .

An **entity-scheme** E is a pair (attr(E), id(E)) , where E is an entity set name, attr(E) is a set of attributes and id(E) is a subset of attr(E) called identifier.

Therefore concrete **entities** e of E can be now defined as tuples on attr(E) .
For a fixed moment of time t the present **entity set** E^t for the entity scheme E is a set of tuples r on attr(E) for which id(E) is a key if id(E) is not empty and
is a multiset (a "set" with duplicates) of tuples r an attr(E) if id(E) is empty.

Given now entity schemes $E_1, \ldots E_k$.

A **relationship scheme** has the form R = (ent(R),attr(R)) where
R is the name of the scheme,
ent(R) is a sequence of entity set names, and
attr(R) is a set of attributes from U .

Given now a relationship scheme $R = ((E_1, \ldots, E_n), \{B_1, \ldots, B_k\})$ and for a given moment t sets $E^t_1, \ldots, E^t_n$.
A **relationship** r is then definable as an element of the cartesian product
$E^t_1 \times \ldots \times E^t_n \times dom(B_1) \times \ldots \times dom(B_k)$.
A relationship set R^t is then a set of relationships, i.e.
$R^t \subseteq E^t_1 \times \ldots \times E^t_n \times dom(B_1) \times \ldots \times dom(B_k)$.

A set $\{E_1, \ldots E_n, R_1, \ldots, R_m\}$ of entity schemes and relationship scheme on a data declaration DD is called **consistent** if the relationship schemes use only the entity schemes $E_1, \ldots, E_n$.

Example 1.6. Let us define for a supermarket database scheme using these notions.
Let U be the set of the following attributes

- Emp (loyees) N(umbe) r
- E (mployees) Address
- D (epartments) Name
- A (rticles) Name
- M (arket) Price
- S (uppliers) Name
- S (uppliers) N (umbe) r
- Emp (loyees) Name
- Salary
- D (epartments) N (umbe) r
- M (arket) N (umbe) r (of the article)
- Quantity
- S (uppliers) Address
- S (uppliers) Price .

The corresponding domains are obvious by the names and therefore omitted.

Given now the following entity schemes

Employees = ({EmpNr, EName, EAddress, Salary}, {EmpNr}),
Department = ({DName, DNr}, {DNr}),
Article = ({AName, MNr, MPrice, Quantity}, {MNr}),
Supplier = ({SName, SAddress}, {SName, SAddress}).

These four kinds of entities cannot exist independent in the supermarket. There are different relationships between these entities. For instance, any employee is working in one department. Any article is sold in at most one department. For each article there exists one supplier which supplies an article by his price and his number. Therefore given now the following relationship schemes

Works-in = ((Employees, Department), ∅),
Manager = ((Employees, Department), ∅),
Sold-In = ((Department, Article), ∅), and
Supplied-by = ((Article,Supplier), {SNr, SPrice}).

The presented relationships have different properties. For each department there exists one and only one manager. Different articles are sold in different departments and an article can be sold in more than one department. Not any employee is a manager. If the same article is sold in different departments then the price is the same.

This information is important for the storage organization, the mapping of this scheme to another database models and therefore needed further.

Given now a set $ERDec = \{E_1, \ldots E_n, R_1, \ldots, R_m\}$ of consistent entity and relationship schemes. Let **R**(ERDec) be the set of all entity and relationship sets $\{(E^t{}_1, \ldots, E^t{}_n, R^t{}_1, \ldots, R^t{}_m) \mid t \geq 0\}$. Then it is possible to define a function C of **integrity constraints** for the set ERDec: $C : \mathbf{R}(ERDec) \longrightarrow \{0,1\}$.
For a given set ERDec of consistent entity and relationship schemes and a function

C of integrity constraints, the pair ERS = (ERDec,C) is called <u>**entity-relationship scheme**</u>. For an entity-relationship scheme ERS = (ERDec,C), an element er from R(ERDec) is called **entity-relationship database** (<u>**ERS-database**</u>) if C(er) = 1 .

In the literature there are defined different special functions of integrity constraints.

Let us define for $R = ((E_1,\ldots,E_k),attr(R))$ and for each i , $1 \leq i \leq k$, the following tuple $comp(R,E_i) = (m,n)$

specifying that in each moment of time a special entity e from E^t_i appears in R^t at least m and at most n times, e.g.

$comp(R,E_i) = (m,n)$ iff for all t , all $e \in E^t_i$

$$m \leq |\{r \in R^t \mid r(E_i) = e\}| \leq n$$

where by $|M|$ is denoted the cardinality of M . If n is unbounded then it is denoted by (m,.).

The complexity function can be generalized for relationship schemes. Given a relationship scheme $R = ((E_1,\ldots,E_n),\{B_1,\ldots,B_k\})$ and a sequence $E'_1\ldots E'_m$ of entity schemes used in R . The complexity constraint

$comp(R,E'_1\ldots E'_m) = (s,p)$ states now that in each moment t the combination of items from the entity set $E^t_1,\ldots,E^t_n$ which are used in the relationship set R^t the combination is used at least s and at most p times, e.g.

$comp(R,E'_1\ldots E'_m) = (s,p)$ iff for all t, all $e'_i \in E'_i$ with

$r(E'_i) = e'_i$ for some $r \in R^t$

$$s \leq |\{ r \in R^t \mid r(E'_i) = e'_i \}| \leq p .$$

<u>Example 1.6.</u> Let us consider Works-in, Manager and Sold-In . We fix the following complexities:

comp(Works-in,Department) = (1,.) ,
comp(Works-in,Employee) = (1,1) ,
comp(Manager,Employee) = (1,1) ,
comp(Manager,Department) = (1,1) ,
comp(Sold-In ,Department) = (0,.) ,
comp(Sold-In ,Article) = (1,.) ,
comp(Supplied-by,Article) = (1,.),
comp(Supplied-by,Supplier) = (1,.) .

This expresses that each employee is working in some department and only there, that each department has at least one employee and generally a lot of employees. The manager-department association is an one-to-one relationship. Each article is sold somewhere. A department is selling generally a lot of articles.

For the case of binary relationships we are able to introduce special kinds of relationships.

Let be $R = ((E_1,E_2),attr(R))$. We say that

if it holds for

R is of type	$comp(R,E_1) \in$	$comp(R,E_2) \in$
1:1	$\{ (0,1) , (1,1) \}$	$\{ (0,1) , (1,1) \}$
1:n	$\{ (l,k \mid l \in \{0,1\}, l \leq k\}$	$\{ (0,1) , (1,1) \}$
1:n	$\{ (l,k \mid l \in \{0,1\}, l \leq k\}$	$\{ (l,k \mid l \in \{0,1\}, l \leq k\}$
n:1	$\{ (0,1) , (1,1) \}$	$\{ (l,k \mid l \in \{0,1\}, l < k \text{ or } l = k\}$

This definition is weaker than the complexity definition but in most cases sufficient. We say that R is an **one-to-one relationship** if it is of type 1:1, that R is an **one-to-many relationship** if it is of type 1:n and not of type 1:1 and that R is a **many-to-many relationship** if it is of type m:n and not of type 1:n nor 1:1 nor n:1 .

This complexity properties are not only properties of relationships. For instance the existence of an employee depends from the existence of a department.
A binary relationship $R = ((E_1,E_2),attr(R))$ is called **hierarchical** if the existence of $e_2 \in E^t{}_2$ depends from the existence of a related $e_1 \in E^t{}_1$.
We can add in our example also a relationship between employees expressing the chief relationship between employees.
A relationship scheme $R = ((E_1,\ldots E_k),\ attr(R))$ is called **recursive** if for different $i,\ j$ $E_i = E_j$.

Example 1.6. Let us delete in the supermarket example the relationship scheme Manager and add the following entity scheme and relationship scheme.
Chief = ({Name, Nr, Phone}, {Nr}),
Is-chief-of = ((Department, Chief) , Ø),
Is-an-employee = ((Chef,Employee), Ø) .
The last relationship scheme is of the following kind
comp(Is-an-employee,Chief) = (1,1) ,
comp(Is-an-employee,Employee) = (0,1) .

This expresses that a chief of a department is also an employee.

Now we consider special kinds of relationships. Given two entity schemes E_1 = (attr(E_1,K_1) , E_2 = (attr(E_2,K_2) and a relationship scheme R = ((E_1,E_2),attr(R)) between them.
R is called **IS-A relationship** (E_1 IS-A E_2) if it is a 1:1 relationship and for each moment of time t holds: Is $e_1 \in E^t_1$ then there exists $e_2 \in E^t_2$ with $e_1(A) = e_2(A)$ for $A \in$ attr(E_1) attr(E_2) .
Therefore the IS-A relationship is a special type of relationship schemes R = ((E_1,E_2),attr(R)) with comp(R,E_1) = (1,1) and comp(R,E_2) = (0,1) .
For $K_1 = \emptyset$, R is called **ID relationship** if it expresses an identification relationship between the entity set of E_1, called weak entity-set, which cannot be identified by its own attributes, but has to be identified by its relationship with the entity set of E_2 .

Now we introduce a graphical representation language for entity-relationship schemes called **entity-relationship diagrams (ERD)** using the following bricks.
Given a data scheme DD = (U,$\underline{D}$,dom) and a set of consistent entity and relationship schemes ERDec = {E_1,...,E_n,R_1,...,R_m} .
The entity-relationship diagram is a finite labeled digraph G_{ERDec} = (U_ERDec,H) where H is the set of directed edges where an edge can be of one of the following forms:
(i) $E_i \longrightarrow A_j$; (ii) $R_i \longrightarrow A_j$; (iii) $R_i \longrightarrow E_j$.
E-Vertices are represented graphically by rectancles, A-Vertices and R-Vertices are represented graphically by circles and diamonds, respectively. If R is a IS-A relationship or an ID relationship then $R \longrightarrow E_1$ is replaced by $R \longleftarrow E_1$.
The edges $R_i \longrightarrow E_j$ are labeled by comp(R_i,E_j) = (n,m) or by 1 if comp(R_i,E_j) $\in$ {(0,1),(1,1)} and by n if
comp(R_i,E_j) $\in$ { (l,k | l$\in${0,1}, l$\leq$k , k > 1} . The edges $E_i \longrightarrow A_j$ can be labeled by dom(A_j) . The identifiers of an entity are underlined.

The following diagrams continue and simplify our previous examples.

Example 1.2.

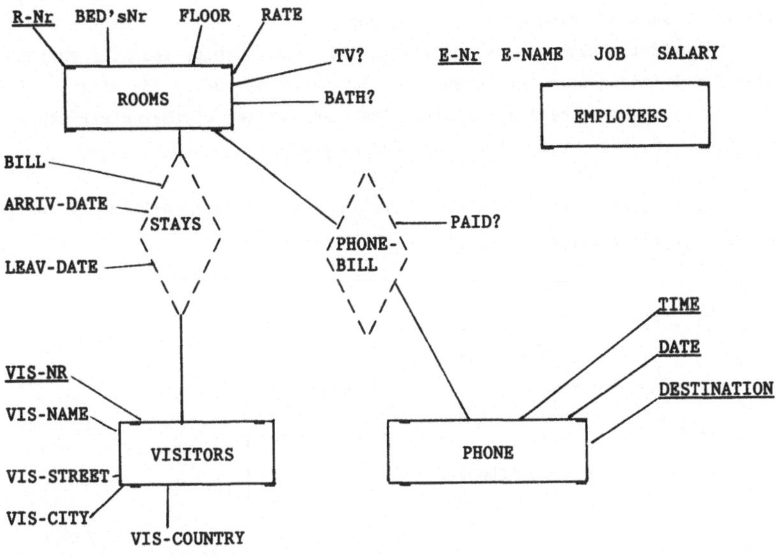

Example 1.3.

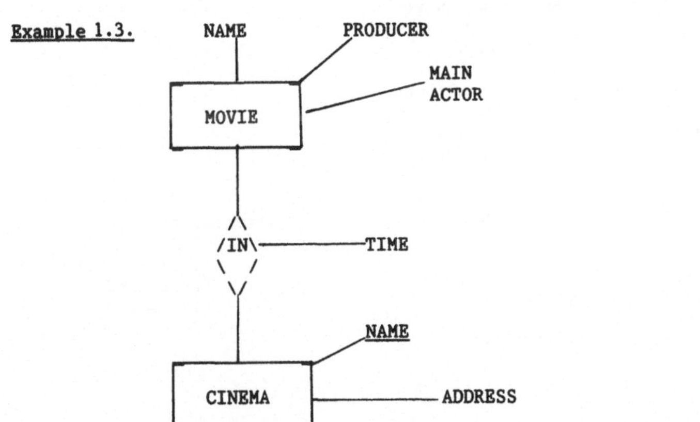

The entity-relationship model is a more general model as the relational model, the hierarchical model and the network model. These three models can be considered as special entity-relationship models.

Obviously, the relational model is an entity-relationship with only entity schemes where the sets of identifiers are not empty.

If we consider only binary and 1:n or 1:1 relationships then the entity-relationship model passes into the **network model**. If additionally the diagram is an ordered set of trees according to increasing complexities of the relationships with roots $E_1,\ldots,E_k$ then we get the **hierarchical model**.

Example 1.7. The following simplified entity-relationship diagram defines a network model for the university database.

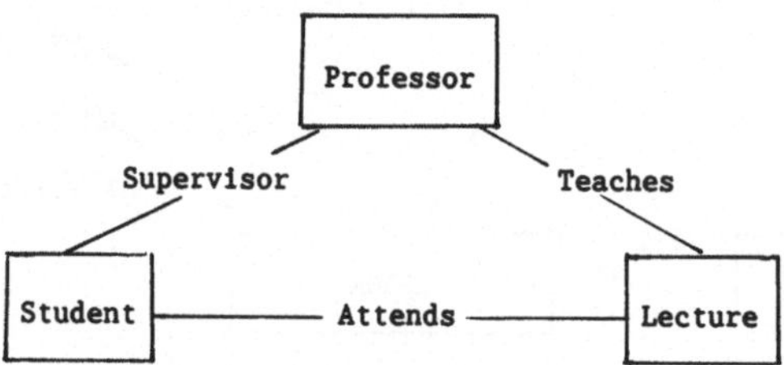

Example 1.8. The following simplified entity-relationship diagram represents a hierarchical model for the university database

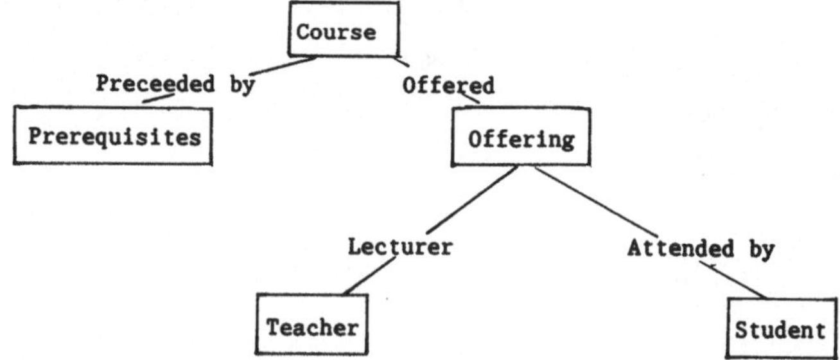

2. THE RELATIONAL ALGEBRA

Many relational queries can be formulated in terms of expressions whose operands represent relations and whose operators are the relational operations. Codd's relational algebra is a high-level language in which questions can be put simply and succinctly /CODD 72/. Concepts from relational algebra have been incorporated into the design of several new database query languages, into view conceptions and into the conception of internal database schemata /IMLI 82/. Expressions in relational algebra manipulate tables of information by means of high-level operations such as select, project, and join. In section 2.1. an algebraic language is introduced. The underlying principle in algebraic languages is to consider the information we wanted to select can be expressed in relations obtained by successive application of database operators. In chapter 2.3, we consider the algebraic dependencies as an important application of the algebraic language.

2.1. THE ALGEBRAIC LANGUAGE

Now there are relations and relational databases, what can be done with them? The content of a database varies with time, so we will consider how to alter a relation. Suppose, we wish to put more information into a database. An "add" operation on the database is performed. We must be able to undo what we do, which calls for a "delete" operation. Instead of adding or deleting an entire tuple or an entire relation, only a part of a tuple or a relation should be modified. Modification can be understood as a binary operation on databases. The relational algebra is a procedural query languages. Query languages are languages in which a user requests information from a database. In the algebraic language called relational algebra, the user instructs the system to perform a sequence of operations on the database to compute the desired result. Many query languages are based on the relational algebra. SQL is one example of such an algebraic query language.
There are five fundamental operations in the relational algebra. These are the projection, the union, the restricted complement, the selection and the extension. The other operations like the intersection, the joins (natural and Theta), the sum, the quotient, and the cartesian product can be defined using the fundamentals operations. It is also possible to choose other operations as the fundamental.

Let us first introduce some set theoretic notions. For sets X, Y ,

the union of sets X and Y is denoted by X u Y or shorter by X Y ,
the intersection of the sets X and Y is denoted by $X \cap Y$,
the difference of these sets is denoted by X - Y .
If X is a subset of Y then this fact will be denoted by $X \subseteq Y$.
For a relation scheme RS = (U , D , dom) where $U = \{ A_1,\dots,A_n\}$, and a set X the set of all tuples on X is denoted by D_X , i.e.

$$D_X = \{ t : X \to \bigcup_{D \in \underline{D}} D \mid t(A) \in dom(A) \} = \{ t_{|X} \mid t \in T(RS)\}.$$

1. Unary and binary operations on one relation scheme.

Given a relation scheme RS = (U,$\underline{D}$,dom) where $U = \{A_1,\dots,A_n\}$.

1.1. The projection

Given a subset X of U and a relation r on RS. The projection of r to X which denoted by r[X] is defined as the set

$$r[X] = \{ t_{|X} \mid t \in r \} .$$

If we represent the relation r as a table, then the operation of its projection over the set of attributes X is interpreted as the selection of those columns of r which correspond to the attributes X and elimination of duplicate rows in a table obtained by such selection.

1.2. The (restricted) complement.

Because of the finiteness of relational databases and the extent of $\underline{D}$ we need a finite operation.
Let us define now the (restricted) complement - r as the set of all tuples which uses values from r but which are not elements of r , i.e.

$$\dot{-}r = \{ t \in T(RS) - r \mid t(A) \in r[A] \text{ for each } A \in U \} .$$

1.3. The union.

Given two relations r , r' on RS. Then the union of r and r' is the set

$$r \cup r' = \{ t \in T(RS) \mid t \in r \text{ or } t \in r' \}.$$

1.4. The intersection.

Given two relations r, r' on RS. Then the intersection of r and r' is the set

$$r \cap r' = \{ t \in T(RS) \mid t \in r \text{ and } t \in r \} .$$

1.5. The difference.

Given two relations r, r' on RS. Then the difference of r and r' is the set $r-r' = \{ t \in T(RS) \mid t \in r \text{ and } t \notin r' \}$.

1.6. The selection.

Let us first define conditions on $\underline{D}$. An atomar condition is a condition of the form $A \theta B$ and $A \theta a$ for $A, B \in U$ $\theta \in \{ =, \neq, <, >, \leq, \geq \}$ and $a \in dom(A)$. Any atomar condition is a condition. Given two conditions α, β then $(\alpha \wedge \beta)$, $(\alpha \vee \beta)$, $\neg\alpha$ are also conditions.

Given a relation r on RS.

For atomar conditions we can now define the selection $\sigma_{\alpha}(r)$ as follows:

$\sigma_{A \theta B}(r) = \{ t \in r \mid t(A) \theta t(B) \}$;

$\sigma_{A \theta a}(r) = \{ t \in r \mid t(A) \theta a \}$.

For conditions α, β the selections $\sigma_{(\alpha \wedge \beta)}$, $\sigma_{(\alpha \vee \beta)}$, $\sigma_{\neg \alpha}$ are defined as follows:

$\sigma_{(\alpha \wedge \beta)}(r) = \sigma_{\alpha}(r) \cap \sigma_{\beta}(r)$;

$\sigma_{(\alpha \vee \beta)}(r) = \sigma_{\alpha}(r) \cup \sigma_{\beta}(r)$;

$\sigma_{\neg\alpha}(r) = - \sigma_{\alpha}(r)$.

For simple selections there can be used also another notation:

$r : (A \theta a) = \sigma_{A \theta a}(r)$;

$r : (A \theta B) \doteq \sigma_{A \theta B}(r)$;

$r : t[X] = \sigma_{A1 = a1 \wedge A2 = a2 \wedge \ldots \wedge Ak = ak}(r)$

for $X = \{A1,\ldots,Ak\}$, $t \in r$, $t[X] = (a1,\ldots,ak)$;

$r:(X=Y) = \sigma_{A1=B1 \wedge \ldots \wedge Ak=Bk}(r)$ where

$X = \{A1,\ldots,Ak\} \subseteq U$, $Y = \{B1,\ldots,Bk\} \subseteq U$ (X,Y)-restriction of r .

For $X = \{A\}$, $Y = \{B\}$ the (X,Y)-restriction is denoted by $r:(A=B)$.

1.7. The anti-projection.

For $X \subseteq U$, $Y = U-X$ the anti-projection on Y of the relation r on RS is a relation with the attribute set Y with tuples for which for any X-value there exists a tuple in r and is noted by $r]Y[$, i.e.

$r]Y[= \{ t_{|Y} \mid t \in r$ and for any $t' \in D_X$ there is in r a tuple t'' with $t''_{|X} = t'$ and $t''_{|Y} = t_{|Y} \}$.

2. Binary operations defined on two relation schemes.

Given now two compatible schemes $RS = (U,\underline{D},dom)$, $RS' = (U',\underline{D'},dom')$.

2.1. <u>The extension of RS to RS+RS'</u>

By RS+RS' is denoted the scheme $(U \cup U', \underline{D} \cup \underline{D'}, dom'')$ with $dom''(A) = dom(A)$ for $A \in U$ and $dom''(A) = dom'(A)$ for $A \in U'$. Given a relation r on RS. The extension Ex(RS,RS')(r) is defined by the set

$Ex(RS,RS')(r) = \{ t \in T(RS+RS') \mid t_{|U} \in r \}$.

2.2. <u>The (natural) join</u>.

Given relations r (on RS) and r' (on RS'). The (natural) join $r * r'$ of r and r' is the set

$r * r' = \{ t \in T(RS+RS') \mid t_{|U} \in r$ and $t_{|U'} \in r' \}$.

Obviously, for RS = RS' the natural join passes into the intersection. For $U \cap U' = \emptyset$ the natural join is the cartesian product. The natural join can be expressed as the intersection of extensions, i.e.

$r * r' = Ex(RS,RS')(r) \cap EX(RS',RS)(r')$.

2.3. <u>The sum</u>.

Given relations r and r' defined on RS and RS'. Then the sum $r + r'$ of these two relations can be defined as the set

$r + r' = Ex(RS,RS')(r) \cup EX(RS',RS)(r')$.

Obviously, for RS = RS' the sum is the ordinary set union.

2.4. <u>The Theta(θ)-join</u>

Given two relations r, r' (on RS and RS'), two attributes $A \in U$, $B \in U'$ and $\theta \in \{<,>,=,\leq,\geq,\neq\}$. The Theta-join of r and r' is defined as the set

$r *(A\ \theta\ B) r' = \{ t \in T(RS+RS') \mid t_{|U} \in r$ and $t_{|U'} \in r'$ and $t(A)\ \theta\ t(B) \}$.

2.5 <u>The quotient</u>

By RS - RS' is denoted the scheme $(U-U', \underline{D}, dom_{|U-U'})$.

The quotient $r \div r'$ (or the division) of two relations r and r' on RS and RS' is used for the evaluation of queries which includes phrases of the form "for all" and is defined for U' with $U' \subseteq U$ as the set

$r \div r' = \{ t \in T(RS-RS') \mid \forall t' \in r' \ \exists t'' \in r : t''_{|U'} = t' \wedge t''_{|U-U'} = t \}$.

Obviously, the quotient can be defined using the following equality

$r \div r = r[U-U'] - ((r[U-U'] * r') - r)[U-U']$.

2.6. <u>The Cartesian product</u>.

If the sets U and U' are disjoint, the join of relations r, r' is called

Cartesian product and noted as $r \times r'$.

<u>Example 1.4.</u> Given the following schemes.
LECTURER = ({lec#,name,category},{set-of-words},dom),
COURSE-UNIT = ({course#,title,lec#},{set-of-words},dom).
Let us consider the following relations r_1, r_2 for LECTURER and COURSE-UNIT.

lec#	name	category
001	Knuth	FProf
002	Wiederhold	AsoProf
003	Gauss	FProf
005	Shennon	AssProf

course#	title	lec#
462	Databases	002
300	Data Structures	001
126	PASCAL 1	004
101	Analysis 1	003
456	Algorithmics	001

Let be now defined

$r_3 = r_1[\{name, category\}]$;

$r_4 = \sigma_{category = FProf}(r_3)$

$r_5 = - \sigma_{course\# > 300} (r_2)$;

$r_6 = \sigma_{lec\# = 001} (r_2) * r_1$;

$r_7 = r_5[\{title\}] \cap r_6[\{title\}]$;

$r_8 = r_4 + r_7$;

$r_9 = r_8 \div r_4 = r_7$.

Then we get the following relations

r_3	name	category
	Knuth	FProf
	Wiederhold	AsoProf
	Gauss	FProf
	Shennon	AssProf

r_4	name	category
	Knuth	FProf
	Gauss	FProf

r_5	course#	title	lec#
	462	Databases	001
	456	Algorithmics	002
	462	Algorithmics	001
	462	Algorithmics	002
	456	Databases	001
	456	Databases	002

r_7	title
	Algorithmics

r_6	course#	title	lec#	name	category
	300	Data Structures	001	Knuth	FProf
	456	Algorithmics	001	Knuth	FProf

r_8	name	category	title
	Knuth	FProf	Algorithmics
	Gauss	FProf	Algorithmics

Some of the operations defined above can be defined in another way. Different other operations can be defined using the above introduced. For instance, we can define a full complement as a set

$r^{-1} = T(RS) - r = \{ t \in T(RS) \mid t \notin r\}$.

If one of the domain sets is infinite the full complement of finite relations generates an infinite relation but the (restricted) complement of a finite relation is finite. That's why the (restricted) complement is only used in databases.

For the definition, some properties of and connections between operations can be used. The operations sum, join and intersection are idempotent, associative and commutative, i.e. for example

$r_1 \cup r_1 = r_1$, $r_1 \cup (r_2 \cup r_3) = (r_1 \cup r_2) \cup (r_3)$, $r_1 \cup r_2 = r_2 \cup r_1$.

Since the definition of the operations is connected with the underlying attribute set the operations Cartesian product and θ-join are associative and commutative, but not idempotent. The complement of a complement of a relation r is a subset of r . Sum and join are double distributive, i.e.

$(r_1 + r_2) * r_3 = (r_1 + r_3) + (r_2 + r_3)$, $r_1 + (r_2 * r_3) = (r_1 + r_2) * (r_1 + r_3)$.

The full complement has the following properties for two relations r_1, r_2 :

$(r_1 + r_2)^{-1} = (r_1^{-1} * r_2^{-1})$; $(r_1^{-1} + r_2^{-1}) = (r_1 * r_2)^{-1}$ (de Morgan's law).

Union and intersection are also double distributive and with the full complement possess de Morgan's law. Unfortunately, the complement does not fulfill these properties. For instance for the relation scheme RS = (U , $\underline{D}$, dom) where U = {A,B} , and the relations $r_1 = \{(0,0),(1,1),(0,1)\}$, $r_2 = \{(0,1),(1,0)\}$, we get $r_1 * r_2 = \{(0,1)\} = -(r_1 * r_2)$, $-(r_1 + r_2) = \emptyset$, $(-(-r_1))*(-r_2) = \emptyset$, but $-((-r_1) + r_2) = \{(0,0)\}$.

For the relation scheme RS = (U , $\underline{D}$, dom) , $X,Y,Z \subseteq U$, and relations r_1 and r_2 on RS, we get

$(r_1[X] * r_2[Y])[Z] = (r_1[X \cap Z] * r_2[Y \cap Z])$ if $X \cap Y \subseteq Z$,

$(r_1[X] \times r_2[Y])[Z] = (r_1[X \cap Z] \times r_2[Y \cap Z])$ if $X \cap Y = \emptyset$,

$(r_1[X] * r_2[Y])[Z] \subseteq r_1[X \cap Z] * r_2 [Y \cap Z]$,

$(r_1[X] \cup r_2[Y])[Z] = r_1[Z] \cup r_2[Z]$ if X = Y.

Given a relation scheme RS = (U , $\underline{D}$, dom) where $U = \{ A_1,\ldots,A_n\}$ and a partition X, Y , Z of U . It is known /THAL 84/ that for a relation r on RS there exist relations r_1 and r_2 with the properties

$r_1[X] = r[X]$, $r_2[Y] = r[Y]$ and $(r_1[XV] * r_2[YV])[XY] = r[XY]$

if $|r[XY]| \leq |D_V|$.

The last property describes the decomposition of a relation r using hidden attributes. If $|V| = 1$ we get the Pawlak database model /PAWL 73/. Furthermore, object-oriented database modeling can be understood as relational database modeling with hidden attributes which are used as object identifiers.

Most of the implementations of relational databases do not include all of these operators. We can limit ourselves to some basic operators using the above listed properties.

Further, it is possible to define the operations using formulas.
The join can be described by the formula $.(P_1(x,z) \wedge P_2(y,z) \text{ ---> } P_3(x,y,z))$.
The projection can be defined by the formula $.(P_1(x,y) \text{ --> } P_2(x))$.
The union $r_3 = r_1 \cup r_2$ is defined by the formula $.(P_1(x) \vee P_2(x) \text{ --> } P_3(x))$.
The intersection $r_3 = r_1 \cap r_2$ is defined by the formula
$.(P_1(x) \wedge P_2(x) \text{ --> } P_3(x))$.
Therefore, the language based on the predicate logic as introduced in chapter 1.1. has at least the expressiveness of the algebraic language. The logical language is even more expressive. For example, the transitive closure r^* of a binary relation r can be expressed thus:
$.(P(x,y) \text{ --> } P^*(x,y))$, $.(P(x,z) \wedge P^*(z,y) \text{ --> } P^*(x,y))$.
It is well known, that this cannot be done in relational algebra /AHUL 79/ and thus this language is indeed more expressive than the relational algebra.

2.2. RELATIONAL EXPRESSIONS

A formal system for reasoning about different kinds of constraints over relational expression can be described. A relational expression is any well formed expression built up from predicate names and relational operators.

A family of formal languages can be defined over relation schemes. Given now compatible relation schemes $RS1 = (U1 , \underline{D}_1 , dom_1)$ where
$U = \{ A_{11},\ldots,A_{1n}\},\ldots,$ $RSl = (U_l , \underline{D}_l , dom_l)$ where $U_l = \{ A_{l1},\ldots,A_{lm}\}$.
Let $DRS = RS1,RS2,\ldots,RSl$. Let U be the union of $U_1,\ldots, U_l$.

A formal language L_{DRS} over DRS comprises the following symbols:
$R_1,\ldots, R_l, c_A$, - , ^ , v , -> , (,) , Pow(U), = , x , u , + ,
where c_A is a constant symbol from a nonempty set of constants for each attribute $A \in U$ and Pow(U) is the set of all subsets of U .

A relational expression of L_{DRS} is inductively defined as follows :
(1) a predicate name R_i is an (atomic) expression over the corresponding

set U_i;
(2) if e is an expression over X and A, A', B $\in$ X , Y $\subseteq$ X , then the projection e[Y] is an expression over Y , and the restriction e:(A=A') and the selection e:(B=c_m) are expressions over X ;
(3) if e and f are expressions over X and Y , then the product (e x f) is an expression over XY if X $\cap$ Y = $\emptyset$, the join (e*f) is an expression over XY, and, if X = Y , the union (e u f) and the difference (e-f) are expressions over X .

A relational expression which is built from one atomic expression R_i by using only the projection and join (in arbitrary order and sequences) is called <u>i-expression</u> .

Using the definition of the operations the set opposed to an expression can be defined for DRS-databases. We are given now a DRS-database $(r_1,\ldots r_l)$. The set $e(r_1,\ldots,r_l)$ can be defined inductively as follows:

(1) if e = R_i then $e(r_1,\ldots,r_l) = r_i$;
(2) if e = e'[Y] then $e(r_1,\ldots,r_l) = (e'(r_1,\ldots,r_l))[Y]$;
 if e = e':(A=A') then $e(r_1,\ldots,r_l) = (e'(r_1,\ldots,r_l))$:(A=A') ;
 if e = e':(B=c_m) then $e(r_1,\ldots,r_l) = (e'(r_1,\ldots,r_l))$:(B=$c_m$) ;
(3) if for # $\in$ {x , * , u , - } e = f#f' then
 $e(r_1,\ldots,r_l) = f(r_1,\ldots,r_l)$ # $f'(r_1,\ldots,r_l)$.

These formal languages can be used also for describing the connections between conceptual and external level in the three level model of database representation. The conceptual level corresponds to a database or relation scheme. The external level corresponds to the view of the whole or a part of the conceptual scheme as would be seen by a group of users concerned with a particular application. The external level can be defined by relational expressions. Another more restrictive possibility for definition of the external level is described by the concept of scheme morphism in /REI 84/. A third definition of the external level using formulas is considered in chapter 3.1.

2.3. ALGEBRAIC DEPENDENCIES

The relational data model is defined as a relational database which satisfies some semantic constraints. Most of these constraints can be formalized and defined as formulas in some logical languages. It is also possible to define these constraints using an algebraic language. Algebraic dependencies are introduced and considered in /YAPA 82/ as a unifying approach to the theory of dependencies. There, algebraic dependencies are introduced for extended schemata with an infinite collection of copies of predicate names and it is shown the equivalence of this class with a later defined class of dependencies, BV-dependencies.

Given a relation scheme $RS = (U, \underline{D}, dom)$ where $U = \{A_1,\ldots,A_n\}$. An <u>algebraic dependency</u> over RS is an assertion of the form $e_1 \subseteq e_2$ where e_1 and e_2 are 1-expressions from L_{RS} over the same set X, $X \subseteq U$.
The two dependencies $e_1 \subseteq e_2$ and $e_2 \subseteq e_1$ together are denoted by $e_1 = e_2$.

An RS-database (r) is called <u>model</u> of the algebraic dependency $e_1 \subseteq e_2$ if $e_1(r) \subseteq e_2(r)$. An algebraic dependency α follows from an algebraic dependency β if any model of β is also a model of α (denoted by $\beta \models \alpha$). This definition can be also extended to sets of algebraic dependencies.

It is not difficult to see that for algebraic dependencies the following assertion is satisfied.

Corollary 2.3.1. Given a relation scheme $RS = (U, \underline{D}, dom)$ where $U = \{A_1,\ldots,A_n\}$. Let e_1, e_2 and e_3 be 1-expressions over X, Y, and Z resp. Any $(RS,\emptyset)$ database (r) is a model of the following algebraic dependencies, i.e. the following algebraic dependencies are valid in any $(RS,\emptyset)$-database (r):

(1) $(e_1[W])[V] = e_1[V]$ for $V \subseteq W \subseteq X$;

(2) $e_1[X] = e_1$;

(3) $e_1 * e_1[W] = e_1$ for $W \subseteq X$;

(4) $(e_1 * e_2)[X] \subseteq e_1$;

(5) $e_1 * e_2 = e_2 * e_1$;

(6) $e_1 * (e_2 * e_3) = (e_1 * e_2) * e_3$;

(7) $(e_1 * e_2)[V W] \subseteq (e_1 * e_2[W])[V W]$ for $V \subseteq X$, $W \subseteq Y$;

(8) $(e_1 * e_2[W])[V W] = (e_1 * e_2)[V W]$ for $V \subseteq X$, $W \subseteq Y$, $X \cap Y \subseteq W$;

(9) $(e_1 * e_2)[W] = (e_1[X \cap (YW)] * e_2[Y \cap (XW)])[W]$;

(10) $e_1[VW] \subseteq (e_1[V] * e_1[W])$ for $V, W \subseteq X$.

The statements (7), (8), and (9), the only one that are not totally trivial, simply states that in the projection one operand of a join may restrict the common attributes of the two operands, and therefore, enrich the result of the join. (8) states that the result of the join remains unaffected if the common attributes are used in later projection. Statement (9) summarizes the statements (7) and (8). Corollary 2.3.1. can be used for the query optimization of algebraic queries.

Corollary 2.3.2. /YAPA 82/ Given a relation scheme $RS = (U, \underline{D}, dom)$ where $U = \{A_1, \ldots, A_n\}$. Let e_1, e_2, e_3 1-expressions over X, X and Z resp. from L_{RS} and $V \subseteq X$.

(1) $e_1 \subseteq e_2 \models e_1[V] \subseteq e_2[V]$.

(2) $e_1 \subseteq e_2 \models e_1{*}e_3 \subseteq e_2{*}e_3$.

Using these corollaries it is possible to define C-sequences $a_1, \ldots a_m$ of algebraic dependencies where C is a set of algebraic dependencies and a_i is an element of C or is a valid algebraic dependency by 2.3.1. or is computed from a_j for $j < i$ by 2.3.2.

From a set C of algebraic dependencies, an algebraic dependency a can be derived if there is a C-sequence $a_1, \ldots, a_m$, a (denoted by $C \vdash a$). Only for a restricted case which will be considered in chapter 3, there is an equivalence between $\vdash$ and $\models$. Using 2.3.1. and 2.3.2. a formal system can be defined (see chapter 3.1.).

Let $e_1 = (R[XY] * ((R[YZ] * ((R[XY] * R[YZ])[XY])) * R[XZ])[YZ])[XZ]$ and $e_2 = (((R[XY] * R[YZ])[XZ] * R[YZ])[XY] * (R[XY] * R[XZ])[XZ])[XZ]$. In /YAPA 82/ for $i, j \in \{1,2,\}$ $i \neq j$ is proved that $e_i \subseteq e_j \models e_j \subseteq e_i$ but not $e_i \subseteq e_j \vdash e_j \subseteq e_i$.

A cover of a set Z is a sequence of sets $X_1, \ldots, X_m$ such that their union $X_1X_2\ldots X_m$ is the set Z. For a relation scheme $RS = (U, \underline{D}, dom)$ where $U = \{A_1, \ldots, A_n\}$ and a cover $X_1, \ldots X_m$ of U the algebraic dependency $R[X_1] * \ldots * R[X_m] \subseteq R$ is called <u>join dependency</u> and denoted by $(X_1, \ldots, X_m)$. Because of (10) of corollary 2.3.1. the join dependency $(X_1, .., X_m)$ is also represented by the algebraic dependency $R[X_1] * \ldots R[X_m] = R$.

3. SOME FUNDAMENTALS OF DEPENDENCY THEORY

This chapter deals with the relationship between logic and relational database theory. The aim of the chapter is to show, by many results published in the literature, how logic can provide a formal support to study classic database problems, and in some cases, how logic can go further, helping first in their comprehension, and then their solution. Logic is just a formal system; many other formal systems have been proposed and applied to databases. In the axiomatic approach, a formal system relies upon an object language, semantics or interpretation of formulas in that language and a proof theory.

Relational database consistency is enforced by integrity constraints which are assertions that databases are compelled to obey. Integrity constraints have been classified according to various criteria. A first classification distinguishes between static constraints which are considered here and characterize valid databases and dynamic constraints imposing restrictions on the possible database transitions which are not considered here because their theory in only in the beginning /VIAN 83/, /THAL 84/. Among static constraints which require the argument of relations to belong to specified domains or dependencies to which this text is devoted. As stated in /ULLM 80/, a fundamental idea concerning integrity constraints is that query languages can be used to express them.

3.1. LOGICAL FUNDAMENTALS OF DEPENDENCY THEORY

Several approaches were made with regard to integrity constraints. Of particular interest are the constraints called data dependencies, or briefly dependencies. Essentially, dependencies are formulas in first-order logic stating, for instance, if some tuples, complying with certain equalities and inequalities, are present in the database, then either some other tuples must also exist in the database or some values in the given tuples must be equal or cannot be equal. Most of papers in dependency theory exclusively deal with various aspects of the implication problem, i.e. the problem of deciding for a given set of dependencies and a dependency whether this set implies the dependency. The reason for the prominence of this problem is that an algorithm for deciding implication of dependencies enables us to decide whether two given sets of dependencies are equivalent or whether a given set of dependencies is redundant or whether for a given set of dependencies an equivalent set of dependencies exists which is better for control and maintenance in real life databases. A solution for the last three problems seems a significant step towards automated database design.

We are given a relation scheme RS = (U , <u>D</u> , dom) where $U = \{A_1,\ldots,A_n\}$, a language L(RS) and a class K of formulas from L(RS) . The <u>implication problem for K</u> is to decide, given $C \subseteq K$, $d \in K$, whether $C \models d$.

Real life databases are inherently finite. When we pay only attention to finite databases we face the finite implication problem which is independent of and different from the implication problem. We say that <u>C finitely implies d</u> (denoted by $C \models_{fin} d$) if $r \models\models C$ entails $r \models\models d$ for every finite relation r on RS (d follows finitely from C). The <u>finite implication problem</u> for a class K of L(RS) formulas is to decide, given $C \subseteq K$ and $d \in K$, whether $C \models_{fin} d$. Clearly, if $C \models d$ then also $C \models_{fin} d$.
These notions can be extended to arbitrary compatible sequences DRS = RS1,...,RSn of relation schemes.

<u>Corollary 3.1.1.</u> /BÖRG 85/ The sets $\{(C,d) \mid C \models d , C \subseteq K, d \in K \}$ and $\{(C,d) \mid C \not\models_{fin} d , C \subseteq K , d \in K \}$ are recursively enumerable for recursive enumerable classes K . If $C \models_{fin} d$ entails $C \models d$ for a recursively enumerable class K , then the implication and the finite implication problem are equivalent and recursively solvable.

B. Trachtenbrot proved /TRA 50/ that the formulas valid in the finite case are not recursively enumerable. Therefore, first-order logic is not recursively axiomatizable in the finite case, and soundness and completeness theorem fails for any logical calculus in the finite case.

An important property of implication is its uniformity in some cases. The implication $\models$ is said to be <u>k-ary</u> for a class K if from $C \models d$ for $C \subseteq K$, $d \in K$ follows the existence of a subset C' of C which has at most k elements such that $C' \models d$.

The finiteness theorem for first-order logic states that if $C \models d$ holds there is also a finite subset C' of C such that $C' \models d$.

Now we introduce formal systems as a formalization of recursive enumerability of implication or finite implication.

Given a class L of objects. By a <u>formal system</u> Γ_L is meant a formal object on L with two components, a subset Ax of L called set of axioms and a set Ru of relations on L called rules of inference or (<u>inference</u>) <u>rules</u> (denoted by Γ_L = (Axioms,Rules)). If Rul is an inference rule and if $(d_1,\ldots,d_n,d) \in$ Rul , then we say that $\langle d_1,\ldots,d_n,d\rangle$ is an application of the rule Rul and that d is a <u>direct consequence</u> of $d_1,\ldots,d_n$ under Rules or Rul. In any application $\langle d_1,\ldots,d_n,d\rangle$ of Rul , the elements $d_1,\ldots,d_n$ are called <u>premises</u> of the application and d is called <u>conclusion</u> of the application. By a <u>derivation</u> from $C \subseteq L$ in Γ_L a finite sequence $d_1,\ldots,d_n$ is meant such that each element d_i is either an axiom of Γ_L or d_i is an element of C or d_i is a direct consequence of one or more earlier elements of the sequence under one of the inference rules of Γ_L . A derivation $d_1,\ldots,d_n$ in Γ_L from $C \subseteq L$ is also called a derivation of its last element d_n , and finally an element d is called <u>derivable</u> in Γ_L from C if there exists a derivation of d in Γ_L from C (denoted by $C \vdash_{\Gamma_L} d$).

Inference rules being usually displayed in the forms of a figure in which a horizontal line is drawn, the premises are written above the line, the conclusion below the line and an application condition after the line:

$$\frac{d_1,d_2,\ldots,d_n}{d} \qquad \text{condition } (d_1,d_2,\ldots,d_n,d)$$

Such formal systems are called <u>Hilbert-type systems</u>.

We are given a set of formal objects and a semantic consequence operation $\models$ in L . The system (L , $\models$) will be said to be a <u>semantic system</u> and the system (L , $\models$, Ax) where Ax is a subset of L will be said to be a <u>semantic theory</u>. The usual consequence operation will be in this text the consequence operation defined in chapter 1.1.

A formal system Γ_L = (Ax,Ru) is said to be <u>sound</u> (w.r.t. (L,$\models$)) if when for $d \in L$, $C \subseteq L$ d is derived in Γ_L from C then d follows from C (w.r.t. (L,$\models$)). Expressing this formally, we have $C \vdash_{\Gamma_L} d$ implies $C \models d$. A formal system Γ_L is said to be <u>complete</u> if for $d \in L$, $C \subseteq L$ when d follows from C then d can be also derived in Γ_L from C , or stated formally $C \models d$ implies $C \vdash_{\Gamma_L} d$.

A semantic system (L , $\models$) is said to be <u>axiomatizable</u> if there exists a sound and complete formal system Γ_L (w.r.t. (L , $\models$) (Γ_L is called an

<u>axiomatization</u> of $(L, \models)$).

A semantic system $(L, \models)$ is said to be <u>finitely axiomatizable</u> if there exists a sound and complete formal system Γ_L (w.r.t. $(L, \models)$) with a finite set of rules and a finite set of axioms.

If we consider the class of relation schemes RS = (U , <u>D</u>, dom) where $U = \{A_1, \dots, A_n\}$ and the languages L(RS) it is possible to distinguish more carefully between axiomatizable semantic systems.

A semantic system $(L, \models)$ is said to be <u>U-bounded axiomatizable</u> if there exists a sound and complete formal system Γ_L (w.r.t. $(L, \models)$) with a U-bounded set of rules and a U-bounded set of axioms.

A formal system $\Gamma_L = (Ax, Ru)$ is said to be k-ary if any rule of Ru has at most k premises.

A semantic system $(L, \models)$ is said to be <u>k-ary axiomatizable</u> if there exists a k-ary sound and complete formal system Γ_L (w.r.t. $(L, \models)$) .

One of the most important properties of databases is summarized in the following

<u>Theorem 3.1.2.</u> /CFP 84/ A semantic system (L , $\models$) is k-ary axiomatizable iff the implication $\models$ is k-ary for L .

If we say that a set K is <u>closed under (k-ary) implication</u> if for every $C \subseteq K$ ($|C| \leq k$) and $C \models d$ implies $d \in K$, then, there is a k-ary complete and sound axiomatization for K iff, whenever $C \subseteq K$ is closed under k-ary implication, then K is closed under implication.

Proof. 1. Assume that there is a k-ary complete and sound formal system Γ_L = (Ax,Ru) . Let C be a subset of L that is closed under k-ary implication. For any $C' \subseteq C$ and $d \in L$ we must show that from $C' \models d$ follows $d \in C$. Since $C' \models d$ we get $C' \vdash_{\Gamma_L} d$. Let $d_1, \dots, d_m$ be a derivation of d from C' , i.e. $d_m = d$. By induction it can be easily shown that $d_i \in C$. If $d_i \in C'$ then $d_i \in C$. If $d_i \in Ax$ then since C is closed under k-ary implication for $k \geq 0$ and therefore $Ax \subseteq C$ it follows $d_i \in C$. If $d_1, \dots, d_i \in C$ and $(d_{i1}, \dots, d_{il}, d_{i+1}) \in Ru'$ for some Ru' of Γ_L with $l \leq k$ by soundness of $_L$ and by k-ary closure of C it follows that $d_{i+1} \in C$. We have shown inductively $d_1, \dots, d_m \in C$ and therefore $d \in C$.

2. Assume that there is no k-ary complete and sound formal system. Now we

shall construct a set C^* that is closed under k-ary implication but is not closed under implication.

Let $Ax = \{ d \mid \ \models d , d \in L\}$ and

$$Ru = \{ \frac{C'}{d} \mid C' \subseteq L , C' \neq \emptyset, d \in L , |C'| \leq k , C' \models d \}$$

Now by assumption $\Gamma_L = (Ax,Ru)$ is not complete but sound. It follows that there is a set $C^+ \subseteq L$ and a formula $d \in L$ such that $C^+ \models d$ and $C^+ \not\vdash_{\Gamma L} d$. Let $C^* = \{ D' \mid C^+ \vdash_{\Gamma L} d'\}$.

Since $d \notin C^+$ and $C^+ \subseteq C^*$ it follows that C^* is not closed under implication. By definition of Γ_L we get C^* is closed under k-ary implication because if for $C' \subseteq C^*$, $d \in L$ with $|C'| \leq k$ it holds that $C' \models d$ then there is a rule

$$\frac{C'}{d}$$

in Γ_L .

A formal system Γ_K is called <u>full</u> (or K-full for a given class K of formulas) if it is sound (or K-sound) and complete (or K-complete) for binary implication. A necessary condition for such systems is that a derivation with elements only from K exists.

<u>Theorem 3.1.3.</u> Given a class K of formulas from L(RS) with a finite number of nonequivalent formulas. The implication problem is solvable if and only if there is a sound and complete formal system for K .

Proof. 1. Suppose the implication is solvable and consider the formal system consisting of one inference rule

$$\{ \frac{d_1,\ldots,d_k}{d} \mid \{d_1,\ldots,d_k\} \models d \} .$$

Obviously, this formal system is sound and complete for K .

2. Suppose, Γ_K is sound and complete for K . Let $C \subseteq K$ and $d \in K$ be given. To decide whether $C \models d$ we list every possible sequence of $d_1,\ldots,d_k \in K$ and check whether it is a derivation of d from C by Γ_K . In as much as there is a finite number of nonequivalent formulas in K , this process must terminate. Hence the implication problem for K is solvable.

As mentioned, relational databases can be seen as finite first-order languages which express exactly the first-order properties of relational databases. The question that arises will first-order logic be sufficient in handling finite structures. What happens to recursive axiomatizability, compactness and other famous theorems on first-order logic in the case of finite structures ?

It is well known (see for example /BØRG 85/) that the formulas valid in the finite case are not recursively enumerable. Tiny fragments of first-order logic are not axiomatizable recursively in the case of finite structures. Summaries of results of that sort can be found in /GURE 76/. The proof of a lot of important theorems in first-order logic use a kind of finiteness argument and the finiteness theorems fails if there are only considered finite structures. We note that Craig's Interpolation theorem, the Weak Definability theorem and the Substructure Preservation theorem fail in the case of finite structures. The proof of the Substructure Preservation theorem is easily relativizable (see for example /GURE 84/), for example for general embedded implicational dependencies (for definition see chapter 3.2.1.).

The introduced database schemes differ from classical predicate logic since they are using different domain sets and are therefore many-sorted.

In chapter 1, RS-relational databases are introduced for many-sorted relation schemes RS = (U , $\underline{D}$, dom) where $U = \{A_1,\dots,A_n\}$. If for A, A' $\in$ U , dom(A) = dom(A') the relation r on Rs can be also defined as a one-sorted relation. Using $D = \bigcup_{A \in U}$ dom(A) the first approach is to introduce one-sorted relation schemes where dom can be understood as an arity function.

There is also a second approach. The set L(DRS) of DRS-formulas can be translated to a set of formulas with one-sorted variables in VAR introducing so called sort predicates $\{P_A \mid A \in U\}$ and sort conditions for atomar formulas:
For the relation scheme RS = (U , $\underline{D}$, dom) where U = {A1,...,An} the formula $P(x_1,\dots,x_n) \in$ L(RS) is replaced by
($P(x_1,\dots,x_n)$ --> $P_{A1}(x_1)$ ^...^ $P_{An}(x_n)$) .
The formula $x_1 = x_2$ for the attributes A1, A2 is replaced by
($P_{A1}(x_1)$ ^ $P_{A2}(x_2)$ --> $x_1=x_2$) .
The set of formulas obtained from L(DRS)-formulas by introducing sort predicates will be denoted by L*(DRS) . Using now databases $(r_1,\dots,r_m)$ on DRS and D where D is the union of all domains we see that the two approaches are semantically equivalent.

It is known /KRKR 67/ that standard one-sorted logic has the same expressive power as many-sorted logic with non-empty sets: for each formula d $\in$ L(DRS) and each database $\underline{r}$ for d in which elements have sorts as defined above there is a one-sorted formula d^* and a database $\underline{r}^*$ for d^* such that d is true in $\underline{r}$ iff d^* is true in $\underline{r}^*$. In one-sorted standard logic we have at hand "universal" variables which are more convenient and which have more expressive power together with sorting predicates.

Often in database theory many-sorted variables are used. This approach is not correct /THAL 84/. Almost all constraints and dependencies dealt with in the literature are strong many-sorted formulas.

Now using, standard results in /CHKE 73/ of first-order logic it is possible to characterize classes of DRS-databases. A class $\underline{R}$ of relations on a scheme RS is said to be axiomatizable by formulas from L(RS) if there exists a set of RS-formulas such that $\underline{R}$ is the class of all models of that set C , i.e. $\underline{R}$ = SAT(C). In /MAVA 85/ a Birkhoff-type characterization of axiomatizable classes of databases is given.

Another application of logic to database theory is the description of connections between external and conceptual level of database representations. The external level corresponds to the view of the whole or a part of the conceptual scheme as would be seen by a group of users concerned by a particular application and being responsible for the implementation of the corresponding user programs. The conceptual level corresponds to the relation scheme as defined in section 1. By a database scheme over a database scheme DRS = RS1,RS2,...,RSl where the schemes RSi = (Ui , $\underline{D}$i, domi) with U = $\{A_{i1},\dots,A_{in}\}$ are given we shall mean any sequence (R1,...,Rk,d1,...,dk) where the Ri are pairwise distinct predicate names and every d_i is a DRS-formula such that

$$Ri(x_1,\dots,x_n) \leftrightarrow d_i(x_1,\dots,x_n) \qquad (d_i \text{ is a connecting formula}).$$

A database scheme can be thought of as a mapping which transforms any DRS-database into an external view of this database. This approach can be extended to the inclusion and equivalence of schemes using results on definability of predicates.

3.2. DEPENDENCIES

The class of dependencies is a class of semantic constraints that are to be satisfied by the database of interest. We are given two database schemes DRSj = RSj1,...,RSjk for j $\in$ {1,2} consisting of relation schemes RSji = (Ui , $\underline{D}$ji, domji) where Ui = $\{A_{i1},\dots,A_{in}\}$. A DRS1-databases $(r_1,\dots,r_k)$ and a DRS2-database $(s_1,\dots,s_k)$ are said to be similar if they have exactly the same rela-

tions, that is $r_i = s_i$ for $1 \leq i \leq k$. A formula d from L(DRS) is said to be <u>domain independent</u> if for all similar databases $\underline{r} = (r_1,..r_k)$, $\underline{s} = (s_1,...s_k)$ (the last is defined on some other scheme) $\underline{r}$ satisfies d if and only s satisfies d .
Remember that a structure $\underline{r}$ satisfies a formula d if there is an interpretation I on $\underline{r}$ such that $\underline{r}$ ||== d[I] .

The aim of this special class is to be able to determine the satisfiability of a formula in a DRS-database by merely taking into consideration the values defined by the relations. We can say that domain-independent formulas guarantee that the elements of a response constitute elementary information actually contained in the relation.

A DRS-database $(r_1,...,r_m)$ is said to be trivial if $|r_i| \leq 1$, for $1 \leq i \leq m$.

Domain-independent formulas which hold in any trivial database are called

<u>d e p e n d e n c y</u>.

The main property of dependencies, the domain independence can be considered as the independence of formulas from the used domains in the database scheme. If we consider only dependencies then the formulas can be considered for a class of languages L(DRS) which are using the same attribute sets, the same predicates but which are independent from the underlying domains. This important property of dependencies states the following

<u>Corollary 3.2.1</u>. Given DRS = RS1,...,RSk where RSi = $(U_i,\underline{D}_i,dom_i)$ for $1 \leq i \leq k$. For dependencies d_1, ..., d_p, d from L(DRS) the following conditions are equivalent:

(1) $\{d_1,...,d_p\}$ |≠ d ;

(2) There exists a DRS'-database $\underline{r} = (r_1,...,r_k)$ with DRS' = RS1',...RSk' and RSi' = $(U_i,\{D\},dom'_i)$ for $1 \leq i \leq k$ for which $\underline{r}$ ||== d_i for $1 \leq i \leq p$ and $\underline{r}$ ||=≠ d.

For instance, the formula $\exists x_1...\exists x_n\ P(x_1,...,x_n,c)$ called in /KOBA 86/ existence constraint is not a dependency.

We now introduce two other classes of formulas which where both characterizing as corresponding to adequate logic formulas for database querying. The first class i that of definite formulas characterized by Kuhns in order to formally represent what he called "reasonable" questions.

For a database scheme DRS = RS1,...,RSk where RSi = (U_i, $\underline{D}_i$, dom_i) for $1 \leq i \leq k$, a DRS-formula d is said to be <u>definite</u> if for any database scheme DRS' = RS1',...,RSk' where RSi' = (U_i, $\underline{D}'_i$, dom_i') for $1 \leq i \leq k$ $dom_i'(A) = dom_i(A) \cup \{c_A\}$ the following are equivalent for DRS-, DRS'-databases $\underline{r}$, $\underline{r}'$ which are similar:

(1) $\underline{r}$ ||== d ;

(2) $\underline{r}'$ ||== d .

The second class is that of safe formulas which was defined by J.D. Ullman /ULLM 80/ in order to characterize those formulas which yields only finite relations on infinite domain sets.

A formula d = d(y_1,..,y_p,c_1,...,c_q) from L(DRS) with constant symbols c_1,...,c_q is <u>safe</u> if for any interpretation I and a DRS-database $\underline{r}$

a) $\underline{r}$ ||== d[I] implies I(y_i) is in DOM(d) for any i where DOM(d) denotes the set of elements corresponding to constant symbols occurring in d together with those occurring in the relations of $\underline{r}$;

b) if $\exists$x d'(x) is a subformula of d then $\underline{r}$ ||== d'[I] implies I(x) is in DOM(d');

c) if $\forall$x d'(x) is a subformula of d then $\underline{r}$ ||=/ d'[I] implies I(x) is in DOM(d').

It is easy to show that any definite formula is a domain-independent formula and vice versa. Any safe formula is definite. But using the following examples it can be shown that there are definite formulas which are not safe:

$P_1(x,y) \wedge \exists z\, (P_2(z) \vee P_3(x,y))$; $\exists x \neg P_1(x) \vee \forall y\, P_1(y)$.

But safe formulas provide the same expressive power as definite formulas. Given any definite formula d $\in$ L(DRS) , there exists a safe formula d' $\in$ L(DRS) such that in any DRS-database $\underline{r}$ $\underline{r}$||== d iff $\underline{r}$ ||== d' /NIDE 83/.

Now we get /DIPE 69/, /VARD 81/

<u>**Theorem 3.2.2**</u>. The set of dependencies from L(DRS) is recursively enumerable iff for DRS = RS1,...RSk k = 1 . For DRS = (U,$\underline{D}$,dom) the set of dependencies from L(DRS) is not recursive.

The decision problem for dependencies is to decide whether a given formula is a dependency. This problem is recursively unsolvable. Based on this theorem, more precisely defined classes of formulas are required for an interpretation of "real world dependency sets" and not only of "real world dependencies".

3.2.1. LOGICAL DEPENDENCIES

Given a database scheme DRS = RS1,...,RSk where RSi = (U_i, $\underline{D}_i$, dom_i) for $1 \leq i \leq k$. Now we define some special kinds of dependencies:

A dependency $d \in L(DRS)$ is called

1. uni-relational dependency if it is built from one predicate P_i , i.e. $d = d(P_i) \in L(RSi)$;
2. many-sorted dependency if $d \in L(DRS')$ for a strong many-sorted database scheme DRS' = RS1',...,RSk' where RSi' = (U_i, $\underline{D}'_i$, dom'_i) for $1 \leq i \leq k$ i.e. $dom'_i(A) \cap dom'_j(B) = \emptyset$, i.e. no variable occurs in two different argument positions of a predicate symbol, and only variables which occur in the same argument position of the predicate can be an argument of an equality formula;
3. general embedded implicational dependency (GEID) if
   $d = \forall y_1 \ldots \forall y_k \exists z_1 \ldots \exists z_l \, (d_1 \wedge \ldots \wedge d_p \rightarrow e_1 \wedge \ldots \wedge e_q)$ where $k,p,q \geq 1$, $0 \leq l$,
   the d_i's and e_j's are atomic formulas $P^m_t(x)$ or $y_m = y_t$,
   at least one d_i is a predicate formula $P^m_t(x)$,
   the set of variables occurring in the d_i's is the same as the set of variables occurring in the predicated d_i's , and is exactly $\{y_1,\ldots,y_k\}$,
   the set of variables occurring in the e_j's contains $\{z_1,\ldots z_l\}$ and is a subset of $\{y_1,\ldots,y_k,z_1,\ldots,z_l\}$;
4. general implicational dependency (GID) if d is a GEID with $l = 0$;
5. inclusion dependency (IND) if d is a many-sorted GEID where $p=q=1$ and d_1 and e_1 are predicate formulas;
6. B(eeri-)V(ardi)-dependency if it is a uni-relational GEID;
7. total BV-dependency if it is a BV-dependency with $l=0$;
8. embedded tuple-generating dependency (ETGD) if it is a BV-dependency in which all e_j's are predicate formulas;
9. tuple-generating dependency (TGD) if it is a ETGD with $l = 0$;
10. embedded template dependency (ETD) if it is a many-sorted ETGD with $q = 1$;
11. template dependency (TD) if it is a ETD with $l = 0$;
12. decomposition dependency (DD) if
   $d = .(P(x_1) \wedge \ldots P(x_p) \rightarrow P(x_0))$ ($x_i = x_{i1},\ldots,x_{in}$ for $0 \leq i \leq p$)
   where for all x_{0j} there is a k $1 \leq k \leq p$, with $x_{0j} = x_{kj}$ and for all i , j ,

$1 \leq i < j \leq p$, and k , $1 \leq k \leq n$, from $x_{ik} = x_{jk}$ follows $x_{ik} = x_{ok}$;

13. embedded multivalued dependency (EMVD) if it is a ETD with $p = 2$;
14. multivalued dependency (MVD) if it is a EMVD and a TD .

A tuple-generating dependency means that if some tuples, meeting certain conditions, exist in the relation, then another tuple must also exist in the relation.

A decomposition dependency means that if some tuples, meeting certain main, more restricted conditions and without hidden conditions, exist in the relation, then another tuple must also exist in the relation.

Another important class of functional associations between attributes can be defined as follows:

We denote by $L^- \subseteq L(DRS)$ the set of DRS-formulas which are not built up from predicate names. This set is called set of generalized equality formulas.

A generalized equality formula $x_{11}=x_{12} \wedge \ldots \wedge x_{k1}=x_{k2}$ is called equality formula.

A dependency $.(d_1 \wedge \ldots \wedge d_m \wedge e \dashrightarrow f) \in L(DRS)$ is called

1. general functional dependency (GFD) if $k,m \geq 1$, the d_i's are predicate formulas and e, f are generalized equality formulas;
2. equality generating dependency (EGD) if it is uni-relational, $k,m \geq 1$, the d_i's are predicate formulas and e, f are equality formulas;
3. generalized functional dependency (GD) if it is a uni-relational, many-sorted GFD with $m = 2$;
4. functional dependency (FD) if it is a EGD which is a GD .

A lot of another dependency classes exists in literature (see for example /DEAD 85/, /THAL 84/, /MAI 83/). As mentioned in /DEAD 85/, in practice, these dependencies are never used to the same extend:

Relative usage frequencies of uni-relational dependencies in practical applications today

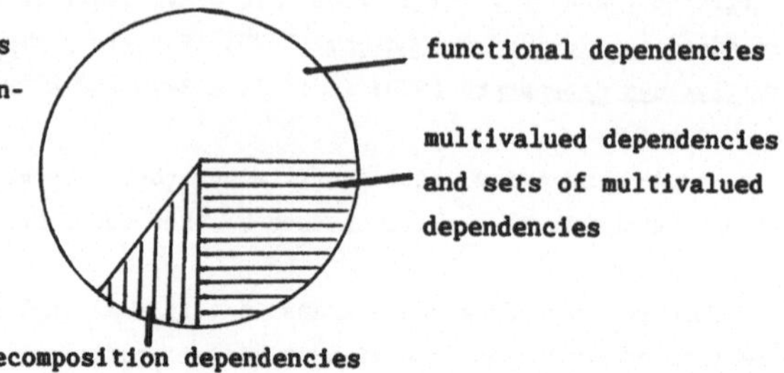

Because of their very easy nature, functional dependencies are by far most widely employed and form the basis for identifying an item.

Overview on some classes of general embedded implicational dependencies

$$d = \forall y_1 \ldots \forall y_k \exists z_1 \ldots \exists z_l \, (d_1 \wedge \ldots \wedge d_p \dashrightarrow e_1 \wedge \ldots \wedge e_q)$$

Conditions for k	l	p	q	Conditions for d_i	Conditions for e_i	Conditions for d	dependency name
		=1	=1	predicates	predicates		inclusion dependency
	=0					uni-relational	BV-dependency
	=0			predicates	predicates	uni-relational	tuple generating depend.
			=1	predicates	predicates	uni-relational many-sorted	embedded template dep.
	=0		=1	predicates	predicates	uni-relational many-sorted	template dependency
	=0			predicates	equalities	uni-relational many-sorted	equality-generating dependency
	=0	=2		predicates	equalities	uni-relational many-sorted	functional dependency
	=0					uni-relational	total BV-dependency
				predicates	predicates	uni-relational	embedded tuple-generating dependencies

At the time of some revision of the book there were introduced some more classes of dependencies most of them remaining out of the scope of this book. But some of the classes are of a high practical importance. The class of closure dependencies /GOSS 89/ seems to be one of those.

Given a relation scheme RS = (U , $\underline{D}$, dom) where U = $\{A_1,\ldots,A_n\}$ and sequences X = $B_1\ldots B_m$ and Y = $C_1\ldots C_m$ of attributes from U where the attributes in the sequences are distinct from each other. The formula

$\forall x_1\ldots\forall x_n\forall y_1\ldots\forall y_n\exists z_1\ldots\exists z_n\ (P(x_1,\ldots,x_n)\wedge P(u_1,\ldots,u_n) \dashrightarrow P(v_1,\ldots,v_n))$

where

$$u_i = \begin{cases} x_j & \text{if } A_j = C_k \text{ and } B_k = A_i \text{ for some } k \\ y_i & \text{otherwise} \end{cases}$$

$$v_i = \begin{cases} x_i & \text{if } A_i = B_k \text{ for some } k \\ y_i & \text{if } A_i = C_k \text{ for some } k \text{ and for no } l\ B_l = A_i \\ z_i & \text{otherwise} \end{cases}$$

is called closure dependency and denoted by X@Y .

Obviously, a relation r on RS satisfies X@Y if for any tuples t , t' from r if $t(C_i) = t'(B_i)$ for i , $1\le i\le m$, then there exists a tuple t" such that $t"(B_i) = t(B_i)$ for i, $1\le i\le m$, and $t"(C_i) = t'(C_i)$ for i, $1\le i\le m$.

The closure dependency can be understood as a constraint which states that the relation is obtained by its transitive closure on X and Y .

For closure dependencies there is necessary only one inference rule for the implication of { X@Y } |= Y@X . It is known that closure dependencies and functional dependencies together have no k-ary axiomatization.

The closure dependency can be generalized to generalized closure dependencies where there is removed the restriction that the attributes in the sequences should be different.

3.2.2. SPECIAL ALGEBRAIC DEPENDENCIES

Now we introduce some special algebraic dependencies for uni-relational databases. The join dependency was already introduced in chapter 2.3.

We are given a relation scheme RS = (U , $\underline{D}$, dom) where U = $\{A_1,\ldots,A_n\}$.

Given a set $\{X_1,\ldots X_m\}$ of subsets of U and $X \subseteq \bigcup_{i=1}^{m} X_i$.

The algebraic dependency $(R[X_1]*...R[X_m])[X] \subseteq R[X]$ is called projected join dependency (PJD) .
As already noticed, the inverse algebraic dependency $R[X] \subseteq (R[X_1]*...*R[X_m])[X]$ is valid in any RS-database $\underline{r}$.

If $\bigcup_{i=1}^{m} X_i = U$ the PJD is called total projected join dependency, and otherwise embedded projected join dependency.

If $X = \bigcup_{i=1}^{m} X_i$ the projected join dependency is called X-join dependency.
If $X \neq U$ the X-join dependency is called embedded join dependency and if $X = U$ the X-join dependency is called join dependency (JD).
The X-join dependency (X_1,X_2) is also called embedded multivalued dependency and denoted by $X_1 \cap X_2 \ ->-> X_1|X_2$ or $(X_1 \cap X_2) ->-> (X_1-X_2)|(X_2-X_1)$.

Join dependencies are shortly denoted by $(X_1,...X_m)$. A join dependency $(X_1,...,X_m)$ is called m-ary join dependency. Let JDEP be the class of all join dependencies and $JDEP_m$ the class of all m-ary join dependencies.

Other kinds of dependencies connected with algebraic dependencies and expressible in special cases with algebraic dependencies are:
inclusion dependency $R_1[X] \subseteq R_2[Y]$ (see chapter 6) ;
transitive dependencies : For $X,Y, Z \subseteq U$, $V = U - XYZ$ and corresponding sequences of variables x, x', y, y', v, v', v", z, z' ,
$\forall x \forall y \forall y' \forall z \forall z' \forall v \forall v' \exists x' \exists v'' (P(x,y,z',v) \wedge P(x,y',z,v') \rightarrow P(x',y,z,v''))$
is called transitive dependency and denoted by $X(Y,Z)$.
If $Y \cap Z = \emptyset$ this dependency is equivalent to $(P[XY]*P[XZ])[YZ] \subseteq P[YZ]$.
extended transitive dependency : For $X_1,...X_p,Y_1,...,Y_q \subseteq U$, the algebraic dependency

$$(\mathop{*}_{i=1}^{p} \mathop{*}_{j=1}^{q} P[X_iY_j])[Y_1...Y_q] \subseteq P[Y_1,...,Y_q]$$

is called extended transitive dependency.

For the set $L = \{ X(Y,Z) \mid X,Y,Z \subseteq U \}$ of transitive dependencies and the implication $\models$ is known /PARE 80/ a sound formal system Γ_{TRD} :
Axioms $XY(Y,Z)$
Rules

$$(1) \quad \frac{X(Y,Z) \ , \ Y(Z,T)}{X(Z,T)} \qquad (2) \quad \frac{X(Y,Z) \ , \ X(T,Z) \ , \ Z(T,YZ)}{X(YT,Z)}$$

$$(3)\ \frac{X(Y,Z)}{X(Z,Y)} \qquad (4)\ \frac{X(YT,Z)}{X(Y,Z)} \qquad (5)\ \frac{X(Y,Z)}{XT(Y,Z)}$$

It is known /DEAD 85/ that there is no complete formal system for transitive dependencies.

It can be denoted that all these algebraic dependencies can be also defined as logical formulas.

Given for a relation scheme RS = (U , D, dom) where $U = \{A_1,\ldots,A_n\}$ a join dependency $d = (X_1,\ldots,X_m)$ and a decomposition dependency

$f = .(P(x_{11},\ldots,x_{1n})\hat{}\ldots\hat{}P(x_{k1},\ldots,x_{kn}) \dashrightarrow P(x_{01},\ldots,x_{0n})) \in L(RS)$.

Given different variables $z_{01},\ldots, z_{0n},\ldots z_{mn}$ from VAR with $z_{ij} \in VAR(A_j)$ (unambiguously with the minimal numbers in VAR) .

Now we define $d_f = (Y_1,\ldots,Y_k)$ with $Y_i = \{A_j \mid x_{ij}=x_{0j}, 1\leq j\leq n\}$, $1\leq i\leq k$, and

$f_d = .(P(u_{11},\ldots u_{1n})\hat{}\ldots\hat{}P(u_{m1},\ldots,u_{mn}) \dashrightarrow P(z_{01},\ldots,z_{0n}))$

$$u_{ij} = \begin{cases} z_{0j} & \text{if } A_j \in X_i \\ z_{ij} & \text{if } A_j \notin X_i \end{cases}$$

Corollary 3.2.3. For any RS-relation r

r ||== d iff r ||== f_d and

r ||== f iff r ||== d_f .

3.2.3. A PROOF PROCEDURE FOR GENERAL IMPLICATIONAL DEPENDENCIES

As a main result of this section, we will characterize the set of all implicational dependencies that is implied by a given set of general implicational dependencies. The characterization yields an algorithm which is related both to the resolution method /CHLE 73/ and the chase method of dependency checking (/ABU79/, /MMS 79/). This procedure is here generalized to general implicational dependencies. It can be extended to general embedded implicational dependencies using the connections between the papers /BEVA 84/ and /GRJA 82/.

We are given a relation scheme RS = (U , D, dom) where $U = \{A_1,\ldots,A_n\}$ and a language L(RS) with variables from VAR .

A substitution of variables is a mapping σ : VAR --> VAR such that if $x \in$ VAR(A) then $\sigma(x) \in$ VAR(A) for all $x \in$VAR .

Given a formula α from L(RS) where

α = $.(\beta \text{ ---> } \beta_1 \wedge \ldots \wedge \beta_m)$. Then this formula is logically equivalent to

$\alpha_1 \wedge \ldots \wedge \alpha_m$ where

$\alpha_i = .(\beta \text{ --> } \beta_i)$ $1 \leq i \leq m$.

Therefore we can assume that the conclusion of α contains a single conjunct, and we write

$\alpha = .(P_1(x_1) \wedge \ldots \wedge P_k(x_k) \text{ ---> } P_O(x_O))$ or

$\alpha = .(P_1(x_1) \wedge \ldots \wedge P_k(x_k) \text{ ---> } y_j = y_i)$.

To state an algorithm, it is required to define a set of atomic formulas $Cl(C, \alpha)$ for a set of GID's and a GID α both with single conjuncts in conclusions by recursion.

Let $\alpha = .(P_{i1}(x_{i1}) \wedge \ldots \wedge P_{im}(x_{im}) \text{ --> } \beta)$.

$Cl_O(C,\alpha) = \{P_{ik}(x_{ik}) \mid 1 \leq k \leq m\}$

$Cl^{\sim}_{k+1}(C,\alpha)$ is got from $Cl_k(C,\alpha)$ by applying the following identification:

if there is a

$\pi = .(P_{11}(u_1) \wedge \ldots \wedge P_{1p}(u_p) \text{ --> } y_i = y_j))$

and a substitution σ such that $P_{1s}(\sigma(u_s)) \in Cl_k(C,\alpha)$

for $1 \leq s \leq p$ then identify $\sigma(y_i)$ and $\sigma(y_j)$ in $Cl_k(C,\alpha)$;

$Cl_{k+1}(C,\alpha)$ = { $P_{1(p+1)}(x)$ | there is a

$\pi = .(P_{11}(u_1) \wedge \ldots \wedge P_{1p}(u_p) \text{ -->} P_{1(p+1)}(u_{p+1}))$ in C and

a substitution σ such that $P_{1s}(\sigma(u_s)) \in Cl^{\sim}_{k+1}(C,\alpha)$, $1 \leq s \leq p$,

and $x = \sigma(u_{p+1})$ }

u $Cl_{k+1}(C,\alpha)$.

$$Cl(C,\alpha) = \bigcup_{k=0}^{\infty} Cl_k(C,\alpha) .$$

Intuitively, $Cl(C,\alpha)$ corresponds to the chase of the tableau /MMS 79/.

It can be proven that

$C \models \alpha$ iff either $\beta \in Cl(C,\alpha)$ for a predicative β

or y_i and y_j are identified in $Cl(C,\alpha)$ for $\beta = y_i{=}y_j$.

Since there is a finite number of atomic formulas composed of α in $Cl(C,\alpha)$ we get that $Cl(C,\alpha)$ can be finitely computed and that there is some k such that $Cl(C,\alpha) = Cl_k(C,\alpha)$.

The computation of $Cl_k(C,\alpha)$ may take up to exponential time in the number of formulas because of the number of substitutions in the generation of each $P_{i(p+1)}(x)$.

There is also another extension of this method. With $Cl(C,\alpha)$ it is possible in the case $C \not\models \alpha$ to construct a model of C which is not a model of α.

The procedure for evaluation of $Cl(C,\alpha)$ is confluent, Church-Rosser, Noetherian, but not effluent in general (for definition see chapter 6.2. or /BÖRG85/).

Using theorem 3.1.3. we get the following property on the existence of sound and complete formal systems.

Corollary 3.2.4. The following classes of dependencies are finitely axiomatizable:
the class of join dependencies and each subclass of this class;
the class of decomposition dependencies and each subclass of this class;
the class of generalized functional dependencies and subclasses.

3.3. TEMPLATE DEPENDENCIES AND TUPLE-GENERATING DEPENDENCIES

In literature, template dependencies are also called total template dependencies, full template dependencies and predicative dependencies. Embedded template dependencies are also called template dependencies.

Since dependencies can be expressed as first-order formulas the relationship between the proof procedure, the chase, presented in chapter 3.2.3. and known proof procedures for first-order logic /CHLE 73/ is not surprising. It turns out that there is indeed, a very strong connection between formal systems for embedded template dependencies and resolution principle and paramodulation. But there are also differences connected with the new quality of many-sorted logic.

The formal systems presented in /CHLE 73/ and /CRAI 67/ are stable w.r.t. derivations within the class of template dependencies. In /BVAR 84/, another independent proof is given for the completeness of some formal system for template dependencies. We present the three formal systems of /BVAR 84/:

Formal system Γ_{TD1} :

Axiom (Ax1) $.(P(x_1)\wedge\ldots\wedge P(x_k) \;{-}{-}{-}{>}\; P(x_i))$, $1\leq i\leq k$

Rules

(P1)
$$\frac{.(P(x_1)\hat{}\ldots\hat{}P(x_m) \dashrightarrow P(x_O))}{S(.(P(x_{p(1)})\hat{}\ldots\hat{}P(x_{p(m)})\dashrightarrow P(x_O)))}$$
for some substitution S , for some permutation p of the permutation group S_m

(P2)
$$\frac{.(P(x_1)\hat{}\ldots\hat{}P(x_m) \dashrightarrow P(x_O))}{S(.(P(x_2)\hat{}\ldots\hat{}P(x_m) \dashrightarrow P(x_O)))}$$
if for some substitution S holds $S(x_1) = S(x_2)$

(P3)
$$\frac{.(P(x_1)\hat{}\ldots\hat{}P(x_m) \dashrightarrow P(x_O)) \text{ , } .(P(y_1)\hat{}\ldots\hat{}P(y_k) \dashrightarrow P(x_1))}{.(P(y_1)\hat{}\ldots\hat{}P(y_k)\hat{}P(x_2)\hat{}\ldots\hat{}P(x_m) \dashrightarrow P(x_O))}$$

Formal system Γ_{TD2} /BVAR 84/

Axiom (Ax1)

Rules (P1)
(P2)

(P4)
$$\frac{\begin{array}{l} .(P(x_1)\hat{}\ldots\hat{}P(x_m) \dashrightarrow P(y_1)) \\ \quad\ldots\ldots\ldots\ldots \\ .(P(x_1)\hat{}\ldots\hat{}P(x_m) \dashrightarrow P(y_p)) \\ .(P(y_1)\hat{}\ldots\hat{}P(y_p) \dashrightarrow P(y_O)) \end{array}}{.(P(x_1)\hat{}\ldots\hat{}P(x_m) \dashrightarrow P(y_O))}$$

Formal system Γ_{TD3} /BVAR 84/

Axiom (Ax1)

Rules (P1)
(P2)

(P5)
$$\frac{\begin{array}{l} .(P(x_1)\hat{}\ldots\hat{}P(x_m) \dashrightarrow P(x_{p+1})) \\ \quad\ldots\ldots \quad \ldots\ldots \\ .(P(x_1)\hat{}\ldots\hat{}P(x_m) \dashrightarrow P(x_m)) \\ .(P(x_1)\hat{}\ldots\hat{}P(x_p)\hat{}P(x_{p+1})\hat{}\ldots\hat{}P(x_m) \dashrightarrow P(x_O)) \end{array}}{.(P(x_1)\hat{}\ldots\hat{}P(x_p) \dashrightarrow P(x_O))}$$

Formal system Γ_{TD4} /BVAR 84/.

Axiom (Ax2) $.(P(x_1) \dashrightarrow P(x_1))$

Rules (P1)
(P2)

$$\frac{\begin{array}{l}.(P(x_{11})\hat{}\ldots\hat{}P(x_{1p}) \dashrightarrow P(y_1))\\ \quad\ldots\qquad\ldots\\ .(P(x_{q1})\hat{}\ldots\hat{}P(x_{qp}) \dashrightarrow P(y_q))\\ .(P(y_1)\hat{}\ldots\hat{}P(y_q) \dashrightarrow P(x_0))\end{array}}{.(P(x_{11})\hat{}\ldots\hat{}P(x_{1p})\hat{}P(x_{21})\hat{}\ldots\hat{}P(x_{qp}) \dashrightarrow P(x_0))} \quad (P6)$$

Theorem 3.3.1. /BVAR 84/, /CRAI 67/ The formal systems Γ_{TD1}, Γ_{TD2}, Γ_{TD3}, and Γ_{TD4} are sound and complete for template dependencies.

Using the following connection, the formal systems presented also be used for derivation of tuple-generating dependencies.
Given a tuple-generating dependency $\alpha = .(P(x_1)\hat{}\ldots P(x_k) \dashrightarrow P(y_1)\hat{}\ldots\hat{}P(y_l))$.
Then for this tuple-generating dependency α there exists a set
$C_\alpha = \{ \ .(P(x_1)\hat{}\ldots\hat{}P(x_k) \dashrightarrow P(y_i) \mid 1\le i\le l \ \}$ of template dependencies.

Corollary 3.3.2. For given tuple-generating dependencies $\alpha 1,\ldots,\ \alpha p,\ \alpha$ the following are equivalent :

(1) $\{ \alpha 1,\ldots,\ \alpha p\} \models \alpha$;

(2) $C_{\alpha 1}$ u...u $C_{\alpha p} \models C_\alpha$;

(3) $C_{\alpha 1}$ u...u $C_{\alpha p} \vdash_{\Gamma_{TDi}} \alpha'$ for any $\alpha' \in C_\alpha$ and some $i \in \{1,2,3,4\}$

It is of interest that the presented formal system can be extended to formal systems for template dependencies and equality-generating dependencies.

Formal system $\Gamma_{TD,EGD}$ /BVAR 84/

Axioms (Ax1)
(Ax3) $.(P(x_1)\hat{}\ldots\hat{}P(x_k) \dashrightarrow x_{ij}=x_{ij})$

Rules (P1)
(P2)
(P4)

$$\frac{.(P(x_1)\hat{}\ldots\hat{}P(x_k) \dashrightarrow x_{ij}=x_{ij})}{.(P(x_1)\hat{}\ldots\hat{}P(x_k)\hat{}P(y_1,\ldots,y_{j-1},x_{ij},y_{j+1},\ldots,y_n) \dashrightarrow P(y_1,\ldots,y_{j-1},x_{ij},y_{j+1},\ldots,y_n))} \quad (P7)$$

$$
\text{(P8)} \quad \frac{.(P(x_1) \wedge \ldots \wedge P(x_k) \dashrightarrow x_{1j}=x_{ij})}{\begin{array}{l} .(P(x_1) \wedge \ldots \wedge P(x_k) \wedge P(y_1,\ldots,y_{j-1},x_{ij},y_{j+1},\ldots,y_n) \dashrightarrow \\ \qquad P(y_1,\ldots,y_{j-1},x_{1j},y_{j+1},\ldots,y_n)\ ) \end{array}}
$$

$$
\text{(P9)} \quad \frac{\begin{array}{l} .(P(x_1) \wedge \ldots \wedge P(x_k) \dashrightarrow P(y_1)) \\ \quad \ldots \qquad \ldots \\ .(P(x_1) \wedge \ldots \wedge P(x_k) \dashrightarrow P(y_m)) \\ .(P(y_1) \wedge \ldots \wedge P(y_m) \dashrightarrow x=y) \end{array}}{.(P(x_1) \wedge \ldots \wedge P(x_k) \dashrightarrow x=y)}
$$

The formal system $\Gamma_{TD,EGD}$ is sound and complete for the class of template and equality-generating dependencies. The rules (P7), (P8) are of special interest implying that the meaning of equalized symbols must be the same.

In /SAUL 82/ a sound and complete formal system for embedded template dependencies is considered.

For the class of template dependencies some properties are known /FMUY 83/. For instance, the TD $\alpha =$
$.(P(x_{11},\ldots,x_{1n}) \wedge \ldots \wedge P(x_{n1},\ldots,x_{nn}) \dashrightarrow P(x_{11},x_{22},\ldots,x_{nn}))$
is the strongest TD in L(RS), i.e. $\{\alpha\} \models \alpha'$ for any TD α' from L(RS).

There exists an infinite sequence of TD's $\alpha_1, \alpha_2, \alpha_3, \ldots$ such that $\{\alpha_{i+1}\} \models \alpha_i$ and $\{\alpha_i\} \not\models \alpha_{i+1}$ for each i. For the construction of such sequences we can use with /FMUY 83/ the following TD's:

$\alpha_i = .(P(x_1) \wedge \ldots \wedge P(x_{p(i)}) \dashrightarrow P(x_0)$ where $p(i) = 2^i$ and

$x_{i1} = x_{(i+2)1}$ for i, $1 \leq i < p(i)-1$, $x_{(p(i)-1)1} = x_{11}$,

$x_{p(i)1} = x_{21}$,

$x_{(2i-1)2} = x_{(2i)2}$ for i, $1 \leq i \leq p(i-1)$,

$x_{(2i)3} = x_{(2i+1)3}$ for i, $1 \leq i < p(i-1)$, $x_{p(i)3} = x_{13}$,

$x_{11} = x_{01}$, $x_{12} = x_{02}$, $x_{23} = x_{03}$.

We can also show that TD's are closed under finite conjunction. That is, we show that if a set of TD's Σ is finite then there is a single TD α that is equivalent to Σ. It is sufficient to prove that for two TD's α_1, α_2 there is an equivalent TD α.

Let $\alpha_1 = .(P(x_1) \wedge \ldots \wedge P(x_m) \dashrightarrow P(x_0))$ and

$\alpha_2 = .(P(y_1) \wedge \ldots \wedge P(y_k) \dashrightarrow P(y_0))$.

Then we define a sequence of the Cartesian product of the variables by

$z_{ij} = (x_{i1},y_{j1}),(x_{i2},y_{j2}),\ldots,(x_{in},y_{jn})$ for $0 \leq i \leq m$, $0 \leq j \leq k$.

Let α =

$.(P(z_{11})\wedge\ldots\wedge P(z_{1k})\wedge\ldots\wedge P(z_{mk}) \;\text{-->}\; P(z_{oo}))$.

Identifying by some substitution $z_{i1},\ldots,z_{ik}$ for any i , $1\leq i\leq m$, we get $\{\alpha\} \models \alpha_1$. Similarly, $\{\alpha\} \models \alpha_2$.

Using (P6) and (P1) we get also $\{\alpha_1, \alpha_2\} \models \alpha$.

<u>Corollary 3.3.3.</u> For any finite set of TD's C there exists an equivalent TD α_C.

3.4. EMBEDDED DEPENDENCIES

For full dependencies, i.e. dependencies of the form .(d) where d is a quantifier-free formula, the implication and axiomatization problems are solvable. For embedded dependencies, i.e. dependencies of the form $\forall x\exists y$(d) where d is a quantifier-free formula, the satisfiability and the finite satisfiability as for $\exists x\forall y\exists x$ - formulas do not coincide, and the corresponding problems are both unsolvable. There exist many kinds of embedded dependencies: embedded multivalued dependencies, first-order hierarchical dependencies, generalized (second-order) hierarchical dependencies, transitive dependencies, generalized transitive dependencies, extended transitive dependencies, crosses, EID, GEID, interrelational dependencies, root dependencies, interdependencies, general dependencies, embedded join dependencies, projected join dependencies, etc.

There are different reasons to introduce embedded dependencies: The feeling of simplicity for embedded multivalued dependencies, the complexity of join dependencies and theorem 3.4.1.

<u>Theorem 3.4.1.</u> For any relation scheme RS = (U , <u>D</u>, dom) , any JD d = $(X_1,\ldots X_m)$ follows from the system $C_d = \{(X_1 , X_2X_3\ldots X_m) , (X_2 , X_3X_4\ldots X_n),\ldots,(X_{m-1},X_m) \}$ of embedded binary join dependencies (which are equivalent to embedded multivalued dependencies α_1, $\alpha_2,\ldots, \alpha_{m-1}$).

Proof. Given an RS-relation r with $r \models\models C_d$. Let $\alpha_1, \alpha_2,\ldots, \alpha_{m-1}$ be the corresponding embedded template dependencies to C_d = $\{(X_1 , X_2X_3\ldots X_m) , (X_2 , X_3X_4\ldots X_n),\ldots,(X_{m-1},X_m) \}$. Let $t_1,\ldots,t_m$ be arbitrary tuples from r with $t_i[X_i \cap X_j] = t_j[X_i \cap X_j]$ for i,j,

$1 \leq i < j \leq m$. If there exists a tuple t in r with $t[X_i] = t_i[X_i]$ then the theorem is proved. We show this by induction. By $r \,||\!\!== \alpha_{m-1}$ there exists in r a tuple t'_{m-1} with $t'_{m-1}[X_{m-1}] = t_{m-1}[X_{m-1}]$ and $t'_{m-1}[X_m] = t_m[X_m]$.
By $r \,||\!\!== \alpha_{m-i}$ for $2 \leq i < m$ there exists in r a tuple t'_{m-i} with $t'_{m-i}[X_{m-i}] = t_{m-i}[X_{m-i}]$ and $t'_{m-i}[X_{m-i+1} \dots X_m] = t'_{m-i+1}[X_{m-i+1} \dots X_m]$.
Now t'_1 is a tuple in r with $t'_1[X_i] = t_i[X_i]$ for i , $1 \leq i \leq m$.

Using the same proof we can also show that for any JD $d = (X_1, \dots, X_m)$ and $C^*_d = \{ (X_i, X'_i) \mid 1 \leq i \leq m ,\ X'_i = \bigcup_{j=1, j \neq i}^{m} X_j \} \cup \{(Y_1, \dots, Y_m)\}$ where
$Y_i = X_i \cap (X_1 \dots X_{i-1} X_{i+1} \dots X_m)$
it holds $C^*_d \models d$.
Using the definition of join dependencies, decomposition dependencies and embedded join dependencies we get $\{ d \} \models C^*_d$.

We remark that for the JD's and EJD's of theorem 3.4.1. the inversion $\{d\} \models C_d$ is not correct.
A relation scheme $RS = (U, \underline{D}, dom)$ is called <u>nontrivial</u> if for all $A \in U$ $|dom(A)| \geq 2$.

<u>Lemma 3.4.2.</u> For any nontrivial relation scheme $RS = (U , \underline{D} , dom)$ there is an RS-relation r that obeys every BV-dependency which is not total, but does not obey some total BV-dependency which is not true in any RS-relation.

Proof. This proof is in the spirit of /FMUY 83/ proof for embedded template dependencies. Let $r = \{0,1\}^n - \{(0,0,\dots,0)\}$. This relation r obeys any (embedded) BV-dependency which is not total. However r violates every nontrivial (i.e. not valid in any relation on RS) which is total.

<u>Corollary 3.4.3.</u> If for a set C of BV-dependencies which are not total BV-dependencies and some total BV-dependency α it holds $C \models \alpha$ then it holds also $\emptyset \models \alpha$ (i.e. $\models \alpha$).

Let EJDEP be the class of embedded join dependencies.

<u>Corollary 3.4.4.</u> If for $d_1, \dots, d_m \in$ EJDEP - JDEP , $d \in$ JDEP , $\{d_1, \dots, d_m\} \models d$ then $(U) \models d$ (i.e. d is trivial).

In theorem 3.4.1., the embedded join dependency $(X_1 , X_2X_3...X_m)$ is a join dependency. The above corollaries show that one join dependency is required for the set C_d in theorem 3.4.1.

In contrast to the general implicational dependencies, the properties of embedded dependencies are substantially unknown. In /SAWE 82/ and /CFP 84/ the following crucial result is shown.

<u>Theorem 3.4.5.</u> The class EMDEP of embedded multivalued dependencies is not finitely axiomatizable.

We prove this theorem using the proof of /CFP 84/.

<u>Lemma 3.4.6.</u> Let $RS = (U , \underline{D}, dom)$ be a relation scheme, $K \subseteq L(RS)$ and let $k \geq 0$ be a constant. Assume that $C \subseteq K$, that $\alpha \in K$, and that
(1) $C \models \alpha$;
(2) if $\alpha' \in C$ then it is not valid that $\{ \alpha' \} \models \alpha$, and
(3) if for $C' \subseteq C$ with $|C'| \leq k$, $C' \models \alpha$ then there is some $\alpha' \in C'$ such that $\{ \alpha' \} \models \alpha$.
Then there is no k-ary axiomatization for K .

Proof. Let $C^* = \{ \alpha \in K \mid \text{there is } \alpha' \in C : \{ \alpha' \} \models \alpha \}$. Since $C \subseteq C^*$ and (2), $\alpha \notin C^*$. Therefore C^* is not closed under implication. We must show that C^* is closed under k-ary implication. Then by theorem 3.1.2. there is no k-ary axiomatization of K .
Now let $C' \subseteq C^*$ with $|C'| \leq k$ and $C' \models \alpha'$ for some arbitrary $\alpha' \in K$. We must show that $\alpha' \in C^*$. For each $\alpha'' \in C'$ let $\beta'' \in C$ such that $\{ \beta'' \} \models \alpha''$. Let $C'' = \{ \beta'' \mid \alpha'' \in C' \}$. Since $C'' \subseteq C'$ and $C' \models \alpha'$ it holds $C'' \models \alpha'$ and by (3) $\alpha' \in C^*$.

<u>Lemma 3.4.7.</u> Given k , $k \geq 0$, there is a relation scheme $RS = (U , \underline{D}, dom)$ such that there is no k-ary axiomatization for embedded multivalued dependencies from L(RS) .

Sketch of the proof of /SAWA 82/. Let be given for a relation scheme $RS = (U , \underline{D}, dom)$ where $U = \{A_0,...,A_{k-1}\}$ the set C and α be defined as

follows: $\{\{A_1\} \to\to \{A_2\} \mid \{A_0\}$, $\{A_2\} \to\to \{A_3\} \mid \{A_0\}$,..., $\{A_{k-2}\} \to\to \{A_{k-1}\} \mid \{A_0\}$,
$\{A_{k-1}\} \to\to \{A_1\} \mid \{A_0\}\ \}$ and
α = $\{A_1\} \to\to \{A_{k-1}\} \mid \{A_0\}$ (C is equivalent to the set of join dependencies $\{(\{A_1,A_2\},\{A_1,A_0\})$, $(\{A_2,A_3\},\{A_2,A_0\})$,..., $(\{A_{k-2},A_{k-1}\},\{A_{k-2},A_0\})$,
$(\{A_{k-1},A_1\},\{A_{k-1},A_0\})$ and α is equivalent to $(\{A_1,A_{k-1}\},\{A_1,A_0\})$).
Then the conditions of lemma 3.4.6. hold.

The proof of theorem 3.4.5. follows from lemma 3.4.7. because EMDEP is a class of formulas with a finite number of nonequivalent formulas.

/SAWA 82/ define a class of subset dependencies which properly contains the embedded multivalued dependencies and which has a finite complete axiomatization for fixed subsets.
A <u>subset dependency</u> (denoted by $Z(X) \subseteq Z(Y)$) is a formula
$\forall x \forall y \forall y' \forall z \forall z' \forall v \forall v' \exists x' \exists v'' (P(x,y,z',v) \wedge P(x,y',z,v') \to P(x',y,z,v''))$
for sequences of variables x, x', y, y', z, z' , v, v', v" which correspond to sets X, Y, Z , V = U - XYZ .

There is for any Z , $Z \subseteq U$, a complete and sound formal system Γ_{SD} :
Axioms ($Ax_{SD,Z}$) $Z(VW) \subseteq Z(V)$ for V,W with $Z \cap (VW) = \emptyset$;
Rule

$$(RU_{SD,Z}) \quad \frac{Z(X) \subseteq Z(Y) \quad , \quad Z(Y) \subseteq Z(W)}{Z(X) \subseteq Z(W)}$$

In comparison with subset dependencies, an embedded multivalued dependency
$X \to\to Y \mid Z$ is a formula
$\forall x \forall y \forall y' \forall z \forall z' \forall v \forall v' \exists v'' (P(x,y,z,v) \wedge P(x,y',z',v') \to P(x,y,z',v''))$
for corresponding sequences of variables.

In connection with theorem 3.1.3. and theorem 3.4.5. the following result is not astonishing.

<u>Theorem 3.4.8.</u> /FAVA 84/, /VARD 84/. The implication problem is unsolvable for the class of embedded template dependencies as well as for GEID's as well as for projected join dependencies.

The smallest superset of EMDEP known to have a complete and sound formal system is the class of ETD /SAUL 82/. Other classes of dependencies, the algebraic dependencies in general case and the GEID which include ETD are also known to be axiomatizable. Theorem 3.4.8. shows the bounds of these axiomatizations.

An embedded join dependency $(X_1,X_2,\dots,X_m)$ for a relation scheme RS = (U , D, dom) where $U = \{A_1,\dots,A_n\}$ is called cross (dependency) if $X_i \cap X_j = \emptyset$ for $1 \le i < j \le m$.

In /BARI 84/ the formal system Γ_{CD} is defined for the relation scheme RS = (U , D, dom) which is sound and complete for crosses.

Formal system Γ_{CD} .

Axiom (X) for $X \subseteq U$;

Rules

(1) $$\frac{(X_1,\dots,X_m)}{(Z_1,\dots,Z_{m'})} \qquad Z \subseteq U,\ Z_i = X_i \cap Z \neq \emptyset,\ 1 \le i \le m';\quad X_i \cap Z = \emptyset \text{ for } i > m'$$

(2) $$\frac{(X_1,\dots,X_m)}{(X_{p(1)},\dots,X_{p(m)})} \qquad \text{for any permutation } p$$

(3) $$\frac{(X_1,X_2,\dots,X_m)}{(X_1X_2,X_3,\dots,X_m)}$$

(4) $$\frac{(X_1X_2,X_3,\dots,X_m)\ ,\ (X_1,X_2)}{(X_1,X_2,\dots,X_m)}$$

A nondecomposition over RS is a subset $X \subseteq U$. A relation r satisfies the nondecomposition X (r¦¦==X) iff it does not satisfy any cross $(X_1,\dots,X_m)$ with $X = X_1X_2\dots X_m$, $X_1 \neq \emptyset$, $X_2 \neq \emptyset$.

Using the transitivity property of nondecompositions (r¦¦==X_1 , r¦¦==X_2 --> r¦¦==X_1X_2) it can be easily shown that Γ_{CD} is sound and complete for cross dependencies /BARI 84/ (for another proof see /PARE 80/).

Given some relation scheme RS = (U , D, dom) . An embedded join dependency $(XY_1,\dots,XY_m)$ is called first-order hierarchical dependency if $Y_i \cap Y_j = \emptyset$ for i,j , $1 \le i < j \le m$, and is denoted by $X : Y_1|Y_2|\dots|Y_m$ /DEAB85/.

It is shown that no finite sound and complete formal system can exist for first-order hierarchical dependencies /PARE 80/. It follows from theorem 3.4.5. using the equivalence of $X : Y_1|Y_2|\dots|Y_m$ and $\{ (XY_1\dots Y_{i-1}Y_{i+1}\dots Y_m, XY_i) \mid 1 \le i \le m \}$.

From the practical viewpoint this means that the closure by successive application of inference rules can not be constructed. Although this is a limitation for the use of algorithms, it is still possible to obtain new dependencies which are of great aid to the user during the conceiving and development phases of a database system. Therefore we present the sound formal system Γ_{FOMD} /DEAB 85/ for the relation scheme RS = (U , <u>D</u> , dom).

Formal system Γ_{FOMD} .

Axioms $\quad X : U-X \quad$ for $X \subseteq U$;

Rules

(1) $$\frac{X: Y_1 | Y_2 | \dots | Y_m}{X: Y_{p(1)} | Y_{p(2)} | \dots | Y_{p(m)}}$$ for some permutation p of {1,2,...,m}

(2) $$\frac{X: Y_1 | Y_2 | \dots | Y_m}{X: Y_1 Y_2 | Y_3 | \dots | Y_m}$$

(3) $$\frac{X: Y_{11} Y_{12} Y_{13} | Y_2 | Y_3 | \dots | Y_m}{XY_{11} : Y_{12} | Y_2 | Y_3 | \dots | Y_m}$$ for $Y_{11} \cap Y_{13} = Y_{11} \cap Y_{12} = Y_{12} \cap Y_{13} = \emptyset$

(4) $$\frac{X : Y_1 | Y_2 Y_3 | Y_4 | \dots | Y_m \ , \ XY_1 : Y_2 | Y_3}{X: Y_1 | Y_2 | Y_3 | \dots | Y_m}$$

(5) $$\frac{XY_{11} : Y_{12} | Y_2 | Y_3 | \dots | Y_m \ , \ X : Z_1 | Z_2}{X : Y_1 \cap Z_1 | Y_1 \cap Z_2 | Y_2 | Y_3 | \dots | Y_m}$$

3.5. GENERAL FUNCTIONAL DEPENDENCIES

The purpose of this chapter is to consider the general functional dependencies which are a type of database dependencies not previously discussed in the literature and to show that a finite axiomatization for different kinds of general functional dependencies does not exist. The meaning of a general functional dependency is that in a relation whenever there are k tuples fullfilling certain properties these k tuples must then also show some other properties. In particular, a generalized functional dependency is a special case with k = 2 . For another special case, the set of equality generating dependencies, there exists an axiomatization.

Remember, that a dependency α from L(RS) is called general functional dependency (short GFD) if $\alpha = .(\alpha_1 \wedge ... \wedge \alpha_k \wedge \beta \dashrightarrow \beta')$ where α_i are predicate formulas and the β, β' are generalized equality formulas from L^-.

A general functional dependency from L(RS) is called many-sorted (or typed) if RS is strong many-sorted (i.e. dom(A) $\cap$ dom(B) = $\emptyset$ for different A,B from U).

A uni-relational general functional dependency $.(P(x_1) \wedge ... \wedge P(x_k) \dashrightarrow \beta)$ is called <u>normalized general functional dependency</u> (short NGFD) if $\beta = \bigvee x_{ij} = x_{ij}$ (β is a disjunction of equalities).

<u>Theorem 3.5.1.</u> A many-sorted uni-relational general functional dependency is equivalent to a set of normalized functional dependencies.

Proof. Given a many-sorted uni-relational GFD

$\alpha = .(P(x_1) \wedge P(x_2) \wedge ... \wedge P(x_k \wedge \alpha' \dashrightarrow \beta)$ where all variables in the sequences $x_i = x_{i1},...,x_{in}$ are different.

From the theory of Boolean functions /JALU 81/ it is known that there are formulas $\alpha_{11},...,\alpha_{1l},...,\alpha_{s1},\beta_{11},...,\beta_{1p},...,\beta_{mp}$ such that $\alpha_{ij} = x_{qv} = x_{ov}$ and $\beta_{ij} = x_{q'v'} = x_{o'v'}$ and $\alpha' \dashrightarrow \beta$ is equivalent to

$((\bigvee_i (\alpha_{i1} \wedge ... \wedge \alpha_{il})) \dashrightarrow (\bigwedge_j (\beta_{j1} \vee ... \vee \beta_{jp})))$.

Therefore, α is equivalent to

$.(\bigwedge_i \bigwedge_j ((P(x_1) \wedge ... \wedge P(x_k) \wedge \alpha_{i1} \wedge ... \wedge \alpha_{il}) \dashrightarrow (\beta_{j1} \vee ... \vee \beta_{jp})))$ and therefore to $.(\bigwedge_i \bigwedge_j (P(x'_{1i}) \wedge ... \wedge P(x'_{ki}) \dashrightarrow (\beta_{j1} \vee ... \vee \beta_{jp}))$ where x_{1i} ,..., x_{ki} is obtained from $x_1,...,x_k$ by identifying the variables according to $\alpha_{i1} \wedge ... \wedge \alpha_{il}$.

We get that for α an equivalent system $\{\alpha_1,...,\alpha_m\}$ of NGFD's exists.

Analogously, the following inversion can be proven.

<u>Theorem 3.5.2.</u> A finite set of normalized general functional dependencies is equivalent to a many-sorted uni-relational general functional dependency.

This theorem can be extended to sets of GFD's.

<u>Example</u>. Let RS and DRS as in example 2 of chapter 1. With NGFD's we can express that any lecture should terminate at most after two terms, i.e. cannot be given longer than for two terms:

$.(P_2(x_1,x_2,x_3,x_4) \wedge P_2(x_1,x'_2,x'_3,x'_4) \wedge P_2(x_1,x''_2,x''_3,x''_4)$
$\quad\quad --> (x_2=x'_2 \vee x_2=x''_2 \vee x'_2=x''_2)\ )$

Instead of implication for the class of many-sorted uni-relational general functional dependencies we can consider the implication for the class of NGFD's. A NGFD $.(P(x_1)\wedge...\wedge P(x_k) --> (\beta_1 \vee...\vee \beta_p))$ is called k-ary.

We are given a relation scheme RS = (U , D , dom) where $U = \{A_1,...,A_n\}$.

Corollary 3.5.3 Suppose that the rule Ru : from $\alpha_1,...,\alpha_m$ infer α_{m+1} is not sound for NGFD's . Let α_{m+1} be a k-ary NGFD . Then there is a RS-relation r with $r||== \{\alpha_1,...,\alpha_m\}$ and $r||=\not\ \alpha_{m+1}$ and $|r| \leq k$.

For the proof of this corollary we consider a RS-relation r with $r||==\{\alpha_1,...,\alpha_m\}$ and $r||=\not\alpha_{m+1}$ which must exist by definition. If r comprises of more than k tuples, then, as explained above, there must be a subrelation r' with k tuples such that for r' the corollary holds.

In /GRMI 85/ numerical dependencies are introduced. A NGFD α =
$.(P(x_1)\wedge...\wedge P(x_k) -->$
$(x_{1i}=x_{2i} \vee x_{1i}=x_{3i} \vee...\vee x_{1i}=x_{ki} \vee x_{2i}=x_{3i} \vee... x_{(k-1)i}=x_{ki}))$
is called k-ary numerical dependency if for some $i_1,...,i_p$ $x_{1j}=x_{1l}$ for $1\leq i\leq k$ and $l \in \{i_1,...,i_p\} \subseteq \{1,...,m\}$ and $x_{1j} \neq x_{1l}$ if $l \notin \{i_1,...,i_p\}$.
For the relation scheme RS = (U , D , dom) where $U = \{A_1,...,A_n\}$ and $X = \{ A_j \in U \mid j \in \{i_1,...,i_p\}\}$ and $B = A_i$ the k-ary numerical dependency α can be denoted by $X --> \langle B\rangle_k$. For k=2 we write $\langle B\rangle_k = \{B\}$.
Obviously, 2-ary numerical dependencies are functional dependencies.

Using theorem 3.1.2, there is shown that there is no finite set of sound and complete rúles for 2-ary and 3-ary numerical dependencies. It follows

Theorem 3.5.4./GRMI 85/ There is no finite sound and complete formal system for numerical dependencies.

The proof is a technical one which uses the impossibility to identify variables of the conclusion of numerical dependencies.

In the literature, numerical dependencies are also called domain dependencies or bounded domain dependencies.

In /KANE 80/ the lossless join problem is considered for numerical dependencies and functional dependencies. The lossless join problem can be formulated as follows: Given a set of dependencies Σ and a join dependency d . Is there a database r satisfying Σ and not d ?
There /KANE 80/ is proven that the lossless join problem is NP-complete if Σ consists of functional dependencies and just one 3-ary numerical dependency.

For other classes of general functional dependencies there exist an axiomatization. For instance, for the class $GEFDEP_m$ of m-ary many-sorted unirelational GFD's a characterization of implication in a k-valued logic can be easily proven.

The important class of equality-generating dependencies has an axiomatization which is equivalent to the paramodulation of /CHLE 73/. Such a formal system is presented in chapter 3.4.

3.6. THE DEDUCTIVE BASIS OF RELATIONS

The idea of using first-order logic in clausal form as a programming language has been applied in many different fields, such as algebraic manipulation, robotics, compilers, and natural language processing. We are of the opinion that a wider use of logic should have a positive effect on the database field, as it provides not only a conceptual framework for formalizing various database concepts, but also a tool for implementing them. It is easy to think of examples in which it is convenient to use general laws to define a relation or a part of a relation. General laws are also useful to avoid redundancy and in connection with updating (trigger concepts). Consider for instance, a relation which is defined in terms of two or more other relations as a view. It is more favorable to state this by general laws than to calculate and to store the relation, explicitly.

The "normalization" of relations is one of the most important tools for database design. The concept of special kinds of dependencies has been proved to be useful in the design and analysis of databases, for instance for normalization. But special kinds of dependencies can be also useful in the reduction of relational databases to the deductive basis. By using special tuple-generating dependencies we get the entry relation from its deductive basis. During the query phase, the rules are used to generate all possible derivations of facts and thereby make them again explicit in the database. But from recursive deduction rules arises the termination problem when the rules are used since potentially, they may lead to infinite derivation paths.

We are given a relation scheme $RS = (U , \underline{D}, dom)$ where $U = \{A_1,\ldots,A_n\}$ and a template dependency $\alpha = .(P(x_1)\wedge..\wedge P(x_k) \rightarrow P(x_0))$ from $L(RS)$. Let r be a relation on RS.

Define the application $\alpha(r)$ of α to r as

$\alpha(r) = r \cup \{ t \mid$ there exist an interpretation I on r such that $I(x_i) \in r$ and $I(x_0) = t \}$.

For the set of template dependencies $C = \{ \alpha_1,\ldots,\alpha_m \} \subseteq L(RS)$ define the application $C(r)$ of C to r as $C(r) = \alpha_1(\alpha_2(\ldots(\alpha_m(r)\ldots))$.

Now $\alpha^k(r)$ denotes the result of k applications of α to r ,

$C^k(r)$ - the result of k applications of C to r ,

$\alpha^*(r)$ - the result of arbitrary many applications of α to r and

$C^*(r)$ - the result of arbitrary many applications of C to r .

These definitions can be easily extended to sequences of relation schemes DRS and to general implicational dependencies

$.(P_1(x_1)\wedge\ldots\wedge P_m(x_m) \rightarrow Q_1(z_1)\wedge\ldots\wedge Q_1(z_1))$

and databases on DRS .

<u>Corollary 3.6.1.</u> For relation schemes $RS = (U,\underline{D},dom)$, a set of template dependencies C and a relation r on RS there exist some k , $k < |r|^{|U|}$, such that $C^*(r) = C^k(r)$.

<u>Corollary 3.6.2.</u> For a relation scheme RS , a set of template dependencies C from $L(RS)$ and a relation r on RS the following are equivalent:

(1) $r \models C$.

(2) $C^*(r) = r$.

Given for a relation scheme $RS = (U ,\underline{D}, dom)$ a set C of template dependencies and a relation r on RS with $r \models C$. A subset r' of r is called C-deductive subset if $C(r') = r$.

A C-deductive subset r' which is minimal , i.e. there is no proper subset r'' of r' such that $C(r'') = r$, is called <u>C-deductive basis</u> of r .

Given a relation r on RS. Let C_r be the set of template dependencies α

with $r \models \alpha$. A C_r-deductive basis of r is called <u>deductive basis</u> of r .

A template dependency $\alpha \in L(RS)$ (or a set of template dependencies $C \subseteq L(RS)$) is <u>bounded</u> iff there exists k such that for any relation r on RS $\alpha^*(r) = \alpha^k(r)$ (resp. $C^*(r) = C^k(r)$). The smallest k with such a property is called the <u>limit</u> of α (resp. C).

Example. Given $RS = (\{1,2,3\},\underline{D},dom)$,
$\alpha = .(P(x_1,x_2,x'_3) \wedge P(x_1x'_2,x_3) \dashrightarrow P(x_1,x_2,x_3))$, $\underline{D} = \{\{0,1\}\}$, and the relation $r = \{(0,0,0),(0,1,1),(0,1,0),(0,0,1),(1,0,0)\}$. The subsets $r' = \{(0,0,0),(0,1,1),(1,0,0)\}$ and $r'' = \{(0,0,1),(0,1,0),(1,0,0)\}$ are α-deductive bases of r . The limit of α is 1 .

The deductive bases of a relation can be also considered as a <u>deductive normal form</u>. These normal forms are more effective according to the storage requirements as the known classical normal forms. Let r be a relation on $RS = (U,\underline{D},dom)$. Let for a multivalued dependency α $r \models \alpha$. Let $d=(X,Y)$ the binary join dependency corresponding to α . Then $r = r[X] * r[Y]$.
We can introduce now a simple complexity measure: $//r// = |r|*|U|$, i.e. length of the tuples multiplied with the number of tuples. Let r' be a α-deductive basis of r. Then we get $//r'// \leq //r//$. There can be found examples where the decomposition using the join dependency α is more effective than deductive normal form. But these examples use the case that $//r[X]// \ll //r[Y]//$. On the other hand, for the set of relations with balanced decompositions (i.e. $//r[X]//$ $//r[Y]//$) deductive normal forms are more effective than the decomposed forms.

There are two main problems.

1. Given a C-deductive basis r of a relation $C^*(r)$. How many steps are required to evaluate $C^*(r)$? What are the estimations of the limit of C ?
2. Given r and C . How to construct a C-deductive basis of r ?

For the second problem there are known some algorithms. The first problem is more difficult. If the set C is unlimited then the utilizing of C-deductive bases is unprofitable.

Example. Given $RS = (U=\{1,2,3,4\},\{ \mathbb{N} \},dom)$ and
$\alpha_1 = .(P(x,y,z,u') \wedge P(x,y,z',u) \dashrightarrow P(x,y,z,u))$,
$\alpha_2 = .(P(x',y,z,u) \wedge P(x,y',z,u) \dashrightarrow P(x,y,z,u))$,

$C = \{ \alpha_1, \alpha_2 \}$

$t_1 = (0,0,0,0)$ and for i, $1 \leq i$,

$t_{2i}[\{1,2,3\}] = t_{2i-1}[\{1,2,3\}]$, $t_{2i}(4) = t_{2i-1}(4) + 1$,

$t'_{2i}[\{1,2,4\}] = t'_{2i-1}[\{1,2,4\}]$, $t'_{2i}(3) = t_{2i-1}(3) + 1$,

$t_{2i+1}[\{1,3,4\}] = t_{2i}[\{1,3,4\}]$, $t_{2i+1}(2) = t_{2i}(2) + 1$,

$t'_{2i+1}[\{2,3,4\}] = t_{2i}[\{2,3,4\}]$, $t'_{2i+1}(1) = t_{2i}(1) + 1$.

Let be $r_1 = \{t_1\}$ and for $i \geq 2$

$r_i = r_{i-1} - \{t_{i-1}\} \cup \{t_i , t'_i\}$, i.e. for example

r_1	r_2	r_3	r_4	r_5	r_6	r_7
0000	0100	0101	0201	0202	0302	0303
	1000	0110	2101	0221	3202	0332
		1000	0110	2101	0221	3202
			1000	0110	2101	0221
				1000	0110	2101
					1000	0110
						1000

Then holds $(0,0,0,0) \notin C^i(r_{i+1})$ and $(0,0,0,0) \notin C^{i-1}(r_{i+1})$ for $i \geq 1$,

i.e. $C^i(r_{i+1}) \neq C^{i-1}(r_{i+1})$.

Therefore C is limited.

<u>Corollary 3.6.3.</u> There exists a set of two multivalued dependencies C (resp. two binary join dependencies) such that C is unlimited. There exists a template dependency α such that α is unlimited.

The last assertion follows for α =

.(P(x,y',z,u')^P(x',y,z,u')^P(x,y',z',u)^P(x',y,z',u) --> P(x,y,z,u))

which implies C in example 2.

A set of decomposition dependencies C is called <u>Sheffer-set</u> if there is a decomposition dependency α_C with $C \models \alpha_C$ and $\{\alpha_C\} \models C$.

Remember, that any finite set of template dependencies has this property. Therefore, the extension of Sheffer-sets to template dependencies is useless.

<u>Corollary 3.6.4.</u> If C is a Sheffer-set with $C \models \alpha_C$ and $\{\alpha_C\} \models C$ for a decomposition dependency α_C then for any relation r on RS it holds

$\alpha_C^*(r) = C^*(r)$.

Theorem 3.6.5./THAL 84/ Given a Sheffer-set C of decomposition dependencies, $C \subseteq L(RS)$. This set C is limited. For any relation r on RS it holds $\alpha_C(r) = C^*(r)$.

For the proof we use the approach of /MINI 83/ to recursive axioms. Given a TD α with the set $Var(\alpha)$ of variables and a subset V of $Var(\alpha)$. A substitution $\sigma\langle x_1 \ldots x_k, y_1 \ldots y_k\rangle = \sigma\langle x_1,y_1\rangle(\sigma\langle x_2,y_2\rangle(\ldots(\sigma\langle x_k,y_k\rangle)\ldots))$ of old variables x_i and corresponding new variables y_i is said to be safe with respect to α and V if $\{y_1,\ldots,y_k\} \cap Var(\alpha) = \emptyset$ and $\{x_1,\ldots,x_k\} \cap V = \emptyset$.
Given two sets of formulas $C_1=\{\beta_1,\ldots,\beta_p\}$ and $C_2 = \{\pi_1,\ldots,\pi_q\}$ with the set V of variables in C_1 and C_2 and the set V_2 of variables used in C_2 . The set C_2 subsumes C_1 w.r.t. V if there is a safe substitution σ w.r.t. $(\pi_1 \wedge \ldots \wedge \pi_q , V_2)$ such that $C_1 \subseteq \{\sigma(\pi_1),\ldots,\sigma(\pi_q)\}$.

Now we define a special sequence $\Omega_i(C,P(x))$ for a set C of TD's and a formula P(x) :

$\Omega_0(C,P(x)) = \{ P(x)\}$;

$\Omega_{i+1}(C,P(x)) = \Omega_i(C,P(x)) - \{ P(y)\} \cup \{P(y_1),\ldots,P(y_m)\}$

for $P(y) \in \Omega_i(C,P(x))$, $.(P(z_1)\wedge\ldots\wedge P(z_m) \dashrightarrow P(z)) \in C$

if there is a safe substitution σ with

$\sigma(P(z)) = P(y)$, and $\sigma(P(z_i))=P(y_i)$.

Any such sequence $\Omega_0(C,P(x))$, $\Omega_1(C,P(x)),\ldots,$ $\Omega_i(C,P(x))$ corresponds to the generation of a new element in $C^i(r)$ and vice versa.

Obviously it holds /CHLE 73/

Lemma 3.6.6. Given a sequence $\Omega_0(C,P(x))$, $\Omega_1(C,P(x)),\ldots,$ $\Omega_i(C,P(x))$,...
If for some j $\Omega_j(C.P(x))$ subsumes $\Omega_{j-1}(C,P(x))$ then the sequence is equivalent to $\Omega_0(C,P(x))$, $\Omega_1(C,P(x)),\ldots,$ $\Omega_{j-1}(C,P(x))$.

Proof of theorem 3.6.5. Given a DD α . Any sequence $\Omega_0(C,P(x))$, $\Omega_1(C,P(x)),\ldots,$ $\Omega_i(C,P(x)),\ldots$ is equivalent to $\Omega_0(C,P(x))$, $\Omega_1(C,P(x))$ since for $\alpha = .(P(x_1)\wedge\ldots\wedge P(x_k) \dashrightarrow P(x_0))$,

$\Omega_1(C,P(x)) = \{P(y_1),\ldots,P(y_k) \}$ and

$\Omega_2(C,P(x)) = \{P(y_1),\ldots,P(y_{i-1}),P(y_{i+1}),\ldots,P(y_k), P(z_1),\ldots,P(z_k)\}$

a safe substitution σ exists for $P(z_1),\ldots,P(z_k)$ such that $\Omega_2(C,P(x))$ subsumes $\Omega_1(C,P(x))$.
With lemma 3.6.6. we get the assertion of theorem 3.6.5.

The next problem is to characterize Sheffer-sets of DD's or of corresponding JD's.
In chapter 5.2. a characterization for Sheffer-sets of binary join dependencies is given. This result can be extended to full hierarchical dependencies as follows.

Theorem 3.6.7. /THAL 84/ Let K be a set of JD's with $X_i \cap X_j = X_i \cap X_k$ for $(X_1,\ldots,X_m) \in K$, $1 \le i \le m$, $1 \le j < k \le m$, $i \ne j$, $i \ne k$.
Then K is a Sheffer-set of JD's iff from
$(X_1,\ldots,X_k) \in K$, $K \models (X_1,\ldots,X_{i-1},Y,X_{i+1},\ldots,X_k)$
follows $K \models (X_1,\ldots,X_{i-1},X_i \cap Y, X_{i+1},\ldots,X_k)$.

3.7. DESIGN BY EXAMPLE

One of the problems plaguing a database designer is the inherent difficulty of extracting from a user the complete semantics of the relations utilized to define the database scheme. Example relations, especially the later described Armstrong relations, can be used as user friendly representation of dependency sets. Different design systems propose the following approach: After the design of the relation scheme the user is asked to present some sample relations. The system extracts dependencies form the presented relations. These dependencies can be used for the decomposition, normalization and representation of relations. This approach is based on the experience that in the average case a considerably small part of a relation suffices for detecting most of the important dependencies which are valid in the database scheme.

Let us introduce the following notions for a database scheme $DS = (RS,C)$ where RS is a relation scheme $(U, \underline{D}, dom)$ with $U = \{A_1,\ldots,A_n\}$. Let C^+ be the set of all dependencies implied by C and let for a class of dependencies K $C^+(K)$ be the intersection of C^+ and K. Let $SAT(C)$ the class of all relations r on the database scheme. Let $K(r) = \{ d \in K \mid r \models d \}$. Obviously, for $r \in SAT(C)$ $C^+ \subseteq K(r)$. For $L(RS)$ and r on RS let $L(r) = \{d \in L(RS) \mid r \models d\}$.

For a given class K of dependencies, <u>design by example</u> means the investigation of relations from $SAT(C)$ in order to discover all the dependencies from K. This design process should be considered as a process of obtaining negative information on the validity of dependencies.

Corollary 3.7.1. For any $r \in SAT(C)$, if $d \notin K(r)$ then $d \notin C^+$.

Normally, a relation is presented tuple by tuple. Therefore, for the design process there is necessary some stability.
A class K of uni-relational dependencies on RS is called input stable if for any relation r on RS and any subset r' of r it holds that $K(r) \subseteq K(r')$. A class K of uni-relational dependencies on RS is called input unstable if there exists a relation r on RS and subsets r' , r" of r such that $K(r') \subsetneq K(r)$ and $K(r) \subsetneq K(r")$.

Corollary 3.7.2. The class of functional dependencies is input stable. The class of equality-generating dependencies is input stable. The class of general functional dependencies is input stable.

Let us consider the following
Example. Given the relation scheme RS = (U , D , dom) where U = {A,B,C} and a relation r = {(0,0,0),(0,1,1),(0,0,1),(0,1,0)} and a subset r' = {0,0,0),(0,1,1)} of r . Obviously, A ->-> B $\in$ MVD(r) for the class MVD of multivalued dependencies, but A ->-> B $\notin$ MVD(r') .

Corollary 3.7.3. The class of multivalued dependencies and any superclass of the class of multivalued dependencies is input unstable.

Therefore for general functional dependencies the stepwise (i.e. tuple-wise) refinement of the set $C^+(K)$ by using sample relations is an appropriate and secure approach. For any class containing at least some multivalued dependencies this approach is not useful.

The efficiency of algorithms generating the set K(r) depends now on the length of the input, i.e. on the number of components in tuples to be considered. Normally, Armstrong or sample relations should use a large number of tuples. Then these algorithms have a higher complexity. Let us consider the cases for which already small subsets r' of r are representative. This assumption would support the strategy of designing by example. Obviously, if the set r' is relatively small in comparison with the set r then we obtain only such dependencies which can be considered as very general. A general learning strategy is based on some assumptions. One of these assumptions could be the assumption that dependencies which are using a smaller number of attributes should be recognized first. The existence of some general functional dependency between attributes from X means that not any X-value can be used. In other words, some X-values are declined. If we obtain the full information on declined values then we know also directly the set of general functional dependencies which is in L(r) . Generally, a subset r' of r is a random subset. Therefore, the information on declined values is random.

For simplicity we consider only the case $\underline{D} = \{\{0,1\}\}$ for the relation scheme RS $= (U, \underline{D}, dom)$ where $U = \{A_1,\ldots,A_n\}$. If r ||== A_1 --> A_2 and $(1,1,\ldots,1) \in r$ then obviously we get $(1,0,x_3,\ldots,x_n) \notin r$ for any $x_i \in \{0,1\}$. Therefore the interval $(1,0,*,*,\ldots,*) = \{(1,0,x_3,\ldots,x_n) \mid x_i \in \{0,1\}\}$ is declined. Let for an interval l be the number of defined elements (rank of the interval). Any interval represents different declination. These declinations can be represented by l implications where l is the rank of the interval. For instance, if the interval $(1,1,0,*,\ldots,*)$ is declined then we get the implications $A_1A_2 \to A_3$, $A_1(-A_3) \to (-A_2)$, $A_2(-A_3) \to (-A_1)$.

Given now a relation r and a subset r' of r . Using r' we obtain an hypothesis on the declined values. The basis of this hypothesis is that the set of declined values obtained using r' is a subset of the set of declined values of r. But it can happen that this set is not sufficient. Therefore, we need the probability of the following statement: A declining interval of rank l is absent in r' but this interval is declined by r.

Now let us consider the probability P(m,n,l) for intervals of rank l on RS with $|U| = n$ and subsets r' with m tuples.

<u>Corollary 3.7.4.</u> $P(m,n,l) \leq \binom{n}{l} 2^l (1 - 2^{-l})^m$.

For the expectation V(m,n,l) of the number of intervals of rank l which have no intersection with the intervals of r' , $P(m,n,l) \leq V(m,n,l)$. Since the number of intervals of rank l is $\binom{n}{l} 2^l$ and the number of orthogonal matrices for an interval of rank l is $2^{mn} (1 - 2^{-l})^m$ we get $V(m,n,l) = \binom{n}{l} 2^l (1 - 2^{-l})^m$.

Using corollary 3.7.4., we get the restrictness of the approach of design by example. The following table represents the maximal number l for the hypothesis on declined intervals for relations r' of length m with n attributes for

W(m,n,l) ≤ 0.01 (P(m,n,l) ≤ 0..01).

n \ m	20	50	100	200	500	1000
10	1	2	3	4	5	6
30	1	2	2	3	4	5
100	1	1	2	3	4	5

Therefore, algorithms which are considering only the properties of the tuples itself require a large number of attributes. In chapters 4 and 5 there are considered excluded constraints. Using the axiomatization of excluded constraints and dependencies presented there there can be developed more effective algorithms.

4. FUNCTIONAL DEPENDENCIES

Dependencies constitute an inherent property of database systems. They express the different ways that data are associated with each other and therefore, the semantics in relational database schemata. Functional dependence is an important property of a relation. In a relation which verifies some functional dependency, there is a functional connection between the parts of tuples. Functional dependencies can be defined like functions $f : X \to Y$ which are mappings satisfying the conditions: 1. For each element $x \in X$ there exists an element $y \in Y$ such that $f(x) = y$.

2. For all $x, x' \in X$: $x = x'$ implies $f(x) = f(x')$.

The second property of functions is used for the definition of functional dependencies. This property can be weakened.

In chapter 4.1., we consider the properties of generalized functional dependencies. In chapter 4.2. functional dependencies are explored. In /DEAD 85/ it is pointed out that functional dependencies constitute nearly 66% of uni-relational dependencies used in practical applications today. In connection with this topic, the design complexity of several problems is considered without neglecting some hard problems. In chapter 4.3, some generalizations of functional dependencies are introduced an contemplated. In a subclass of functional dependencies, the keys are one of the most important constraints. An attribute of a group of attributes may be used to qualify a tuple of a relation. In chapter 4.4 we present some results on the complexity and the structure of sets of keys. The concept of Armstrong databases considered in chapter 4.5. for generalized functional dependencies is of interest in the relational database theory and in mathematical logic and is a fascinating topic which has been studied explicitly for only a few years. This topic is also connected with chapter 3.7. The axiomatization for generalized functional dependencies is used to find an axiomatization for the class of functional and degenerated multivalued dependencies in chapter 4.6.

Let $RS = (U , \underline{D}, dom)$ where $U = \{A_1,\ldots,A_n\}$ be a fixed relation scheme. Let $\underline{D} = \{ \mathbb{N}\}$ (the set of natural numbers with zero).
In this part only RS databases are considered and therefore n, $\underline{D}$, RS are often omitted. We use now the algebraic definition of relations.

4.1. PROPERTIES OF GENERALIZED FUNCTIONAL DEPENDENCIES

In chapter 3.2., generalized functional dependencies are introduced. In this chapter, we shall see that Boolean algebra offers a particularly interesting framework to resolve an essential part of problems dealing with dependencies. This makes available the familiar tools of truth-tables, Karnaugh maps, and syntactic derivations to decide if a given functional dependency is a consequence of some set of generalized functional dependencies. In /SDPF 81/ the family of Boolean dependencies called here generalized functional constraints is introduced. These constraints extend functional dependencies by allowing arbitrary Boolean combinations of attributes. Al-Fedaghi introduced independently a similar notion, the notion of propositional dependencies which is to be considered at the end of this chapter. In this chapter, we consider a subclass of Boolean dependencies, the class of generalized functional dependencies for which the consequence relation is equivalent to the consequence relation for propositional logic. Generalized functional dependencies are equivalent to positive Boolean dependencies /BEBL 85/. Generalized functional dependencies are of importance for a more natural definition of dependencies of functional kind and unifies all these dependencies. They can be introduced in a more intuitive manner.

A pair (f,g) of n-ary Boolean functions is called <u>generalized functional constraint</u>.
Given a relation r on RS with $U = \{A_1,\ldots,A_n\}$.
For a Boolean function f we can define a binary relation $\sim_f$ on r :
$t \sim_f t'$ iff $f(\sigma_1(t,t'),\ldots,\sigma_n(t,t')) = 1$ where $\sigma_i(t,t)$ denotes the function

$$\sigma_i(t,t') = \begin{cases} 0 & \text{if } t(A_i) \neq t'(A_i) \\ 1 & \text{if } t(A_i) = t'(A_i) \end{cases} \qquad 1 \leq i \leq n$$

Now we can define the validity of (f,g) in r :
$r \models (f,g)$ iff for any $t,t' \in r$ from $t \sim_f t'$ follows $t \sim_g t'$.
By $\sigma(t,t')$ let us denote the sequence $\sigma_1(t,t'),\ldots,\sigma_n(t,t')$.

Given a pair (f,g) of n-ary Boolean functions. (f,g) is called <u>generalized functional dependency</u> if $f(1,..,1) \leq g(1,\ldots,1)$.

<u>Corollary 4.1.1.</u> If for some functional constraint and a non-empty relation r

$r \models (f,g)$ holds, then (f,g) is a generalized functional dependency.

Therefore, a generalized functional constraint is a dependency if and only if it is a functional dependency.

Let us first verify that this notation and the notation introduced in chapter 3.2. mean the same.
A generalized equality formula $x_{11}=x_{12} \wedge \ldots \wedge x_{k1}=x_{k2}$ is called equality formula.
A dependency $.(d_1 \wedge \ldots \wedge d_m \wedge e \rightarrow e') \in L(RS)$ is called
generalized functional dependency (GFD) if $k,m \geq 1$, the d_i's are predicate formulas and e, e' are generalized equality formulas and if $m = 2$.
Dependencies α_1, α_2 are called **equivalent** if in any relation r they both are valid in r or they both are false in r.
Obviously, any generalized equality formula α defines a Boolean function
f_α. We define for $\sigma_1,\ldots,\sigma_n \in \{0,1\}$, $d_j = P(x_{j1},\ldots,x_{jn})$, $j \in \{1,2\}$
$f_\alpha(\sigma_1,\ldots,\sigma_n) = 1$ iff $\models \alpha[I]$ with $I(x_{1i}=x_{2i}) = \sigma_i$.
From the theory of Boolean functions we get that for any Boolean function f there are generalized equality formulas α with $f = f_\alpha$. For instance,

$$\alpha_f = \bigvee_{\substack{(\sigma_1,\ldots,\sigma_n) \in \{0,1\}^n \\ f(\sigma_1,\ldots,\sigma_n) = 1}} \bigwedge_{i=1}^{n} \alpha_i^{\sigma_i}$$

for

$$\alpha_i^{\sigma} = \begin{cases} x_{1i} = x_{2i} & \text{if } \sigma = 1 \\ \neg\, x_{1i} = x_{2i} & \text{if } \sigma = 0 \end{cases} .$$

Now we get for any uni-relational GFD

$$\alpha^* = .(P(x_{11},..,x_{1n}) \wedge P(x_{21},\ldots,x_{2n}) \wedge \alpha(x_{11},\ldots,x_{1n}x_{21},\ldots,x_{2n}) \rightarrow \beta(x_{11},\ldots,x_{1n},x_{21},\ldots,x_{2n}))$$

there is some functional constraint (f_α,f_β) with $r \models (f_\alpha,f_\beta)$
iff $r \models \alpha^*$ for any relation r on RS.
From corollary 4.1.1. follows that any generalized dependency is explicitly defined by a GFD. Therefore, we can use the two notions of chapter 3.2.2 and chapter 4.1. similarly. It should be noticed that for generalized functional dependencies there can be defined also directly generalized equality formulas equivalent to the given generalized functional dependency. It is well-known [31] that each Boolean function can be represented by a disjunctive normal form. Therefore the pair (f,g) can be represented by two disjunctive normal forms d_f, d_g. An implication $A \rightarrow B$ of two propositional formulas can be represented by the for-

mulas $\neg A \vee B$ and therefore by a propositional formula $d_{(f,g)}$. From the other hand side, for each propositional formula d there exists a Boolean function f_d such that d and $(1,f_d)$ are equivalent dependencies where by 1 is denoted the Boolean function identically equal to 1 .

<u>Lemma 1</u>. For Boolean functions f,g with $f(1,\dots,1) \leq g(1,\dots,1)$ and a propositional dependency d the following equivalences are valid:
1. (f,g) is equivalent to $d_{(f,g)}$.
2. A is equivalent to $(1,f_d)$.

The proof of this lemma is obvious because of the semantics of general functional dependencies and propositional formulas.

Some special generalized functional dependencies are the strong functional dependency, dual functional dependency, weak functional dependency, monotone functional dependency and key dependency /DEGY 81/, /THAL 85/. The theory of all these special general functional dependencies can be unified and simplified by a theory of general functional dependencies which is based on the following theorem. By $S^n{}_1$ the class of m-ary disjunctions is denoted $(0 \leq m \leq n)$, by $P^n{}_1$ the class of m-ary conjunctions is denoted $(0 \leq m \leq n)$, by $A^n{}_1$ the class of m-ary monotone functions is denoted and by 1 is denoted the tautology. These special subclasses can be expressed by a simpler set of formulas in the language

$\{X \dashrightarrow^{\pi} Y \mid X,Y \subseteq U,\ \pi \in \{F,D,S,W\}\} + \{\underline{X} \dashrightarrow^{M} \underline{Y} \mid \underline{X},\underline{Y} \subseteq Pow(U)\}$.

For $U = \{A_1,\dots,A_n\}$, $\circ,\bullet \in \{\wedge,\vee\}$, $f = x_{i1} \circ\dots\circ x_{im}$, $g = x_{j1} \bullet\dots\bullet x_{jp}$ the general functional dependency (f,g) can be denoted by

$\{A_{i1},\dots,A_{im}\} \dashrightarrow^{\pi} \{A_{j1},\dots,A_{jp}\}$ with

$$\pi = \begin{cases} W & \text{if } \circ = \wedge,\ \bullet = \vee \quad \text{weak functional dependency} \\ D & \text{if } \circ = \vee,\ \bullet = \vee \quad \text{dual functional dependency} \\ S & \text{if } \circ = \vee,\ \bullet = \wedge \quad \text{strong functional dependency} \\ F & \text{if } \circ = \wedge,\ \bullet = \wedge \quad \text{functional dependency} \\ & \qquad\qquad\qquad\ \text{(F normally omitted)} \end{cases}$$.

Analogously, monotone functional dependencies can be expressed by general functional dependencies. Let for $X = \{A_{i1},\dots,A_{im}\} \subseteq U$ f_X be the function $x_{i1} \wedge\dots\wedge x_{im}$ and for $\underline{X} = \{X1,\dots,Xk\}$ $f_{\underline{X}}$ be the function $f_{X1} \vee\dots\vee f_{Xk}$. The monotone functional dependency $\underline{X} \dashrightarrow \underline{Y}$ can be denoted by $(f_{\underline{X}},f_{\underline{Y}})$.
If we consider these subclasses we will use these denotations similarly. By these equivalent expressions, it is possible to use equivalent formulations instead of

the introduced definition of validity of general functional dependencies,

$r \,||\!\!== X \dashrightarrow^{D} Y$ if for any two tuples $t, t' \in r$ if for some $A \in X$ $t(A)=t'(A)$ then for some $B \in Y$ $t(B)=t'(B)$;

$r \,||\!\!== X \dashrightarrow^{W} Y$ if for any two tuples $t, t' \in r$ if for all $A \in X$ $t(A)=t'(A)$ then for some $B \in Y$ $t(B)=t'(B)$;

$r \,||\!\!== X \dashrightarrow^{S} Y$ if for any two tuples $t, t' \in r$ if for some $A \in X$ $t(A)=t'(A)$ then for all $B \in Y$ $t(B)=t'(B)$;

$r \,||\!\!== X \dashrightarrow Y$ if for any two tuples $t, t' \in r$ if for all $A \in X$ $t(A)=t'(A)$ then for all $B \in Y$ $t(B)=t'(B)$;

$r \,||\!\!== \underline{X} \dashrightarrow^{M} \underline{Y}$ if for any two tuples $t, t' \in r$ and for some $X \in \underline{X}$ if for all $A \in X$ $t(A)=t'(A)$ then for some $Y \in \underline{Y}$ and for all $B \in Y$ $t(B)=t'(B)$.

By S_1^n the class of (m-ary) disjunctions is denoted ($m \geq 0$, $m \leq n$),
by P_1^n the class of (m-ary) conjunctions is denoted ($m \geq 0$, $m \leq n$), and
by A_1^n the class of (m-ary) monotone functions is denoted.

generalized functional dependencies	f from	g from	class denoted by	dependency denoted by
strong functional dependencies	S_1^n	P_1^n	SFDEP	$X \dashrightarrow^{S} Y$
dual functional dependency	S_1^n	S_1^n	DFDEP	$X \dashrightarrow^{D} Y$
weak functional dependency	P_1^n	S_1^n	WFDEP	$X \dashrightarrow^{W} Y$
functional dependency	P_1^n	P_1^n	FDEP	$X \dashrightarrow Y$
monotone functional dependency	A_1^n	A_1^n	MFDEP	$X \dashrightarrow^{M} Y$
key dependency	P_1^n	1	KFDEP	$X \dashrightarrow U$

In literature (/DEGY 81/, /BEBL 85/, /THAL 84/), some special applications ofthe class GFDEP of generalized functional dependencies are presented.

Example. Consider the incidence structure of n points and m blocks, each block being a set of points. Let the points be labeled by $A_1, \dots, A_n$. We consider each of the m blocks as a function t_i, $1 \leq i \leq m$, with domain $U = \{A_1, \dots, A_n\}$ where $t_i(A_j) = (i-1)m + j$ if A_j is not in the ith block and $t_i(A_j) = 0$ otherwise. If r is the set $\{t_1, \dots, t_m\}$ then some familiar combinatorial restrictions on incidence structures can be expressed using generalized functional dependencies. For example $r \,||\!\!== \emptyset \dashrightarrow^{W} U$ is equivalent to the condition that any two blocks intersect in at least one point. More generally, let $1 \leq k \leq n$ and let S_k denote the family of all k-element subsets of U . The condition that r represents a graph of n edges and m vertices is expressed by

$r \mid\mid== S_2 \dashrightarrow^M U$, $r \mid\mid== \emptyset \dashrightarrow^M S_1$. Further $r \mid\mid== \emptyset \dashrightarrow^M S_k$ is equivalent to the condition that any two blocks intersect in at least k points. It is of interest that graphical dependencies (see chapter 5) and other join dependencies can be so considered.

Example. We consider a relation TIMETABLE on U ={LECTURER, COURSE-UNIT,STUDENT,CLASSROOM,TIME} with the following restrictions:

1. Any student can at most participate in one course at the same time.
2. Any lecturer gives at most one lecture at the same time.
3. Any classroom is reserved only for one group at the the same time.
4. If there is a lecture given by more than one lecturer then participants are different.

The relation TIMETABLE is given by the following table.

LECTURER	COURSE-UNIT	STUDENT	CLASSROOM	TIME
Smith	Analysis	John	A	Mo-1
Smith	Data Bases	Ali	A	Mo-2
Davis	Systems	John	B	Mo-2
Davis	Analysis	Ali	B	Tu-2
Davis	Algebra	John	A	Tu-1
Asser	Logic	John	A	Tu-2
Asser	Calculus	John	A	We-1
Asser	Systems	Ali	B	Mo-1
Asser	Data Bases	Bob	B	Tu-1
Church	Set Theory	Bob	A	We-2
Beth	Computation	John	A	Th-1
Beth	Computation	Ali	A	Th-2
Carnap	Semantics	John	B	Th-2
Carnap	Semantics	Ali	B	Th-1

These restrictions are represented by the general functional dependencies (f_1,g_1), (f_2,g_2), (f_3,g_3), (f_4,g_4) for $f_1 = x_3 \wedge x_5$, $f_2 = x_1 \wedge x_5$, $f_3 = x_4$, $f_4 = -x_1 \wedge x_2$, $g_1 = x_2$, $g_2 = x_2$, $g_3 = x_3 \vee -x_5$, $g_4 = x_3$. Obviously, the dependency $(x_2 \wedge x_3,\ x_1)$ also holds in the relation. This dependency follows from the introduced dependencies.

Remember, for $k \geq 0$, $\underline{R}_k$ denotes the set of all relations on RS that have at most k tuples. For a class $\underline{R}'$ of relations on RS , also $C \models_{\underline{R}'} \alpha$ is the natural relativization of $C \models \alpha$ to relations in $\underline{R}'$.

<u>Theorem 4.1.2.</u> For any superset $\underline{R}'$ of $\underline{R}_2$, any set C of generalized functional dependencies and a generalized functional dependency d the following are equivalent:

1. $C \models d$.
2. $C \models_{\underline{R}'} d$.

Proof. Let us first for the set $\underline{R}' = \underline{R}_2$ of all two-element relations r with card(r) = 2 prove theorem 4.1.2. For one- or zero-element relations, any dependency is valid. Therefore such relations are not needed to be considered. The direction 1. ==> 2. is trivial. Let now C and d such that $C \not\models d$. Then by definition there exists a relation r in $\underline{R}$ such that $r \models\!= C$ and $r \not\models\!= d$. Therefore there exists a subset r' of r containing two tuples such that $r' \models\!= C$. Because of $r \models\!= C$ it holds also $r' \models\!= C$. Therefore we get $C \not\models_{\underline{R}'} d$.
For arbitrary R' the theorem follows analogously.

This theorem is the basis for the algorithm SATISFIES given below.
<u>Algorithm 4.1.3.</u> SATISFIES
Input: A relation r and a generalized functional dependency (f,g) ;
Output: "true" , if r satisfies (f,g) , "false" otherwise .

SATISFIES(r,(f,g))
If each set of tuples t,t' from r with $f(\sigma(t,t')) = 1$ has g-equal values (i.e. $g(\sigma(t,t')) = 1$ or $t =_g t'$) , return "true" .
Otherwise, return "false".

The algorithm presented above is the same for the case of functional dependencies. Therefore, the dependency satisfaction for generalized functional dependencies is not more complicated than for functional dependencies.

Note that theorem 4.1.2. can be extended for fixed k to sets $(\alpha_1 \wedge \ldots \wedge \alpha_k \wedge \beta \rightarrow \beta')$ of general functional dependencies with k predicate formulas in the premise and $\underline{R}_k$, respectively. For this extension, we can use therefore the k-valued logic.

Now we shall prove the main characterization theorem for implications of generalized functional dependencies.
For Boolean functions f, g the inequality $f \leq g$ holds if for any value tuple σ from $f(\sigma) = 1$ follows $g(\sigma) = 1$.
For a set $S = \{(f_1,g_1),\ldots, (f_m,g_m)\}$ of generalized functional dependencies by ^S is denoted the conjunction $(f_1 \to g_1)\wedge\ldots\wedge(f_m \to g_m)$ of implications of those functions.

__Theorem 4.1.4.__ Let $S = \{(f_1,g_1),\ldots, (f_m,g_m)\}$ and (f,g) be a set of generalized functional dependencies and a generalized functional dependency. Then $\{(f_1,g_1),\ldots, (f_m,g_m)\} \models (f,g)$ holds iff $\wedge S \leq (f \to g)$ holds.

Proof. 1. We prove the theorem first for $m = 1$.
1.1. If $f_1 \to g_1 \not\leq f \to g$ then there exists a value σ with $f(\sigma) = 1$, $g(\sigma) = 0$
and $f_1(\sigma) = 0$ or $f_1(\sigma) = g_1(\sigma) = 1$.
Then we get $r \models f_1 \to g_1$ and $r \not\models f \to g$ for $r = \{ (\sigma), (1,1,\ldots,1)\}$.
Therefore $f_1 \to g_1 \not\models f \to g$.
1.2. Let $r = \{t,t'\}$ a relation from $\underline{R}_2$ with $r \models f_1 \to g_1$ and $r \not\models f \to g$.
Then we get for $\sigma = \sigma(t,t')$
$f_1(\sigma) = g(\sigma) = 0 \neq f(\sigma)$ or
$f_1(\sigma) = g_1(\sigma) = f(\sigma) = 1 \neq g(\sigma)$.
Thus $f_1 \to g_1 \not\leq f \to g$.
2. The proof of the theorem for $m = 2$ is analogous.
3. From 2. we get that exists for $C = \{(f_1,g_1),\ldots,(f_m,g_m)\}$ a system $C' = \{f_1,g_1),\ldots,(f_{m-2},g_{m-2}),(f'_{m-1},g'_{m-1})\}$ with $C \models C'$ and $C' \models C$. That implies that a functional dependency (f_C,g_C) exists for C equivalent to C .

Theorem 4.1.4. can be proven also in another interesting approach. Remember that by SAT((f,g)) is denoted the set $\{ r \mid r \models (f,g) \}$ (analogous SAT({f,g}) and for sets of GD's C , C' SAT(C C') = SAT(C) $\cap$ SAT(C')). By definition $C \models (f,g)$ iff SAT(C) $\subseteq$ SAT((f,g)) . Then we need for the proof of theorem 4.1.4. the property that the set of GD's is Armstrong (see also chapter 4.5.).

Now we demonstrate the strength of theorem 4.1.4. by a series of intermediate corollaries.

Corollary 4.1.5. Let be (f_1,g_1), (f_2,g_2) generalized functional dependencies.

1. If $f_2 \leq f_1$, $g_1 \leq g_2$ then $\{(f_1,g_1)\} \models (f_2,g_2)$.
2. $\{(f_1,g_1),(f_2,g_2)\} \models (f_1 \wedge f_2,\ g_1 \wedge g_2)$. (conjunction of GD's)
3. $\{(f_1,g_1),(f_2,g_2)\} \models (f_1 \vee f_2,\ g_1 \vee g_2)$. (disjunction of GD's)
4. If $g_1 \leq f_2$ then $\{(f_1,g_1),(f_2,g_2)\} \models (f_1,g_2)$. (generalized transitivity)
5. If $f_1 \leq g_1$ then $\emptyset \models (f_1,g_1)$.
6. $\{(f_1,g_1)\} \models (-g_1,-f_1)$ where for a Boolean function f the negation of f is denoted by - f .

Corollary 4.1.6. For each set of generalized functional dependencies there exists an equivalent general functional dependency.

An example of (f_C,g_C) is the C-root

$$\left(\bigvee_{(f,g)\in C} (f \wedge -g) ,\ \bigwedge_{(f,g)\in C} (-f \vee g) \right)$$

A system C of GD's is called independent if for any $(f,g) \in C$ $C - \{(f,g)\} \not\models (f,g)$ holds.

From corollary 4.1.5. and the denotation $[C] = \{(f,g) \in GFDEP \mid C \models (f,g)\}$ for systems C of GD's we get

Corollary 4.1.7. For any set C of GD's, there exists a number k , $0 \leq k < 2^n$, such that $|[C]| = 3^k 4^m$ with $m = 2^n-k-1$ and $|C| \leq k$ if C is independent.
For any k , $0 \leq k < 2^n$, there exists an independent system C of GD's with $|C| = k$ and $|[C]| = 3^k 4^m$ for $m = 2^n-k-1$.

Corollary 4.1.8. Let be $h(y_1,\ldots y_m)$ an m-ary monotone Boolean function and $(f_1,g_1),\ldots,(f_m,g_m)$ GD's. It holds
$\{(f_1,g_1),\ldots,(f_m,g_m)\} \models (h(f_1,\ldots,f_m) , h(g_1,\ldots,g_m))$.

Corollary 4.1.9. For any system C of GD's there exists an equivalent system C' of weak functional dependencies.

A set of GD's C is called <u>closed</u> if $C = [C]$.

We can introduce a semiorder $\geq$ in GFDEP and maximal elements of closed sets: For GD's (f_1, g_1) , (f_2, g_2) $(f_1, g_1) \geq (f_2, g_2)$ if $f_2 \leq f_1$ and $g_1 \geq g_2$. For a closed set C , a Boolean functions f', g' let be now defined

$$max_C(f') = \bigwedge_{(f,g)\in C} g \quad , \quad min_C(g') = \bigvee_{(f,g)\in C} f \quad ,$$

$$min_C(f') = \bigvee_{(f,g)\in C} g \quad , \quad max_C(g') = \bigwedge_{(f,g)\in C,} f \quad ,$$

$min(C) = \{ (f,g) \in C \mid g = min_C(f') , max_C(g') , (f',g') \in C\}$,

and $max(C) = \{ (f,g) \in C \mid g = max_C(f) , f = min_C(g) \}$.

<u>Corollary 4.1.10.</u> Let C be a closed set of generalized functional dependencies.

1. The structure $(max(C), + , \cap)$ is a distributive lattice for the operations $+$, $\cap$ with

$(f_1, g_1) + (f_2, g_2) = (min_C(g_1 \vee g_2) , g_1 \vee g_2)$,

$(f_1, g_1) \cap (f_2, g_2) = (f_1 \wedge f_2 , max_C(f_1 \wedge f_2))$.

2. A generalized functional dependency (f,g) is an element of C iff there exists an element (f',g') in $max(C)$ such that $f \leq f'$ and $g' \leq g$ holds.

3. For any element (f,g) of $max(C)$, there exists exactly one presentation $(f_1, g_1) + (f_2, g_2) + \ldots + (f_k, g_k)$ with +-irreducible elements of $max(C)$.

In /VTHI 84/ there is proved a stronger result for closure operations. The generalized functional dependency (f,g) is an element of the closed set C iff there are GD's $(f',g') \in max(C)$ and $(f'',g'') \in min(C)$ such that $f'' \leq f \leq f'$ and $g' \leq g \leq g''$.

Now we get using the previous corollaries

<u>Corollary 4.1.11.</u> Any system of pairwise nonequivalent subsets of GFDEP consists of at most $2^{2^n - 1}$ elements. There exists a system of pairwise nonequivalent subsets of GFDEP with exactly $2^{2^n - 1}$ elements.

<u>Corollary 4.1.12.</u>. Testing whether two sets of general functional dependencies are

equivalent is NP-complete. Testing whether two sets of general functional dependencies implies the same set of key dependencies (keys) is NP-complete.

<u>Corollary 4.1.13.</u> Let C be a set of GD's and $X \subseteq U$. The following are equivalent:

(i) $C \models X \dashrightarrow U$.

(ii) $\bigwedge_{A_i \in X} -x_i \leq \bigwedge_{(f,g) \in C} (-f \vee g)$.

(iii) $\bigvee_{(f,g) \in C} (f \wedge -g) \leq \bigvee_{A_i \in X} x_i$.

Numerous algorithms concerning relational databases use a <u>cover</u> for a set of functional dependencies as all or part of their input. Examples are Beeri and Bernstein's synthesis algorithm and the tableau modification algorithm of Aho et al /DEAB 85/. the performance of these algorithms may depend on both the number of functional dependencies in the cover and the total size of the cover. Starting with a smaller cover will make such algorithms faster. In /THAL 84/ several kinds of minimality for covers are defined and, using these corollaries and the theory of covers of Boolean functions /JALU 80/, some basic results of the theory of covers in GFDEP are presented. These results emphasize the importance of the class of functional dependencies for database design.

In /ALTH 88/ there is considered a dependency similar to generalized functional dependencies which could be understood as the representation of generalized functional dependencies by formulas.
Given a set of attributes $U = \{A_1,\ldots,A_n\}$. With each attribute A there is associated a <u>propositional variable</u> A' . For two different tuples t, t' on U the propositional variable A' denotes the proposition : "The two tuples agree in the A-value". The negation of A' , ¬ A' , denotes the contrary, that these tuples have different A -values. Without any loss of generality we denote by A the attribute and the propositional variable.
Given furthermore a set { ^ , v , ¬ , -> , <-> } of logical connectives (conjunction, disjunction, negation, implication, equivalence). Using these connections and the set U there can be defined a set L(U) of <u>propositions</u> or <u>propositional dependencies</u> on U :

1. Any propositional variable is a proposition.
2. If H and H' are propositions then ¬H , (H ^ H'), (H v H'), (H -> H'), (H <-> H') are propositions.

For any pair of different tuples (t,t') and the set L(U) there can be defined an <u>interpretation</u> of propositions:

1. The propositional variable A is said to be valid for (t,t'), if t(A) = t'(A) and otherwise false.

2. ¬H is valid for (t,t') if H is false for (t,t'). (H ^ H') ((H v H'), (H -> H'), (H <-> H')) is said to be valid for (t,t') if H and H' (H or H', ¬H or H', (H -> H') and (H' -> H) respectively) are valid for (t,t').

The validity of H for different t,t' is denoted by (t,t') ¦¦== H .

For sets of attributes $X = \{B_1,\ldots,B_m\}$ the set X is also to be used to denote the proposition $B_1 \wedge \ldots \wedge B_m$.

The notion (t,t')¦¦== H can be extended to r¦¦== H as follows:

The proposition H is valid in r (denoted by r¦¦== H) iff for any pair of different tuples (t,t') from r (t,t') ¦¦== H .

A set <u>H</u> of propositional dependencies is valid in r (denoted by r ¦¦== <u>H</u>) if any element of <u>H</u> is valid in r.

For a subset <u>R</u>' of <u>R</u>, a given set <u>H</u> of propositional dependencies and a propositional dependency we say that <u>H</u> imply H if for any relation r from <u>R</u>' in which <u>H</u> is valid r¦¦== H (denoted by <u>H</u> $\models_{R'}$ H or by <u>H</u> ¦= H for R' = R).

<u>Corollary 4.1.14.</u> For any relation r with ¦r¦ ≤ 1 and any propositional dependency H r ¦¦== H .

Therefore propositional dependencies are dependencies.

<u>Corollary 4.1.15.</u> For any system of propositional dependencies there exists an equivalent propositional dependency. For any propositional dependency there exists an equivalent generalized functional dependency.

<u>Example.</u> The propositional dependency (¬X v Y v ¬Z) denotes the fact that a given relation r satisfies the dependency if for every two different tuples, the X-values or the Z-values differ or the Y-values matches.

Suppose that X Y = U . For a functional dependency X -> Y ,e.g. X is the <u>key</u> of U, the equivalent propositional dependency is ¬X . That is, for any two tuples in the relations on U , the two tuples differ in the X-value.

The above presented example illustrates that two propositional formulas have the same meaning on a given universe U because of the definition of the interpretation: H and the formula H ^ ¬U. The disjunct ¬U is overflowing because of relations are defined to be sets and two tuples of a relation should be different. Therefore the disjunct ¬U can be eliminated in all propositional dependencies or can be added to all propositional dependencies. Instead of considering the whole propositional logic L(U) we add to all dependency sets $\underline{H}$ the axiom $(\neg A_1 \vee \ldots \vee \neg A_n)$ as an axiom to our propositional logic called **dependency propositional logic, DPL.**

Let us denote by ⊧ the consequence relation for dependency propositional logic.

__Theorem 4.1.16__. For a given set $\underline{H}$ of propositional dependencies and a proposition dependency H the following are equivalent:

1. $\underline{H} \models H$.
2. $\underline{H} \mid= H$.

Proof. Obviously in dependency propositional logic the formula ¬U is added to each formula. But this corresponds to the introduced notion of interpretations of propositional formulas. Therefore the proof of the theorem is evident.

Several advantages may be gained by adopting generalized functional dependencies instead of functional dependencies. While generalized functional dependencies are richer in terms of expressing additional constraints in the world of two-tuple relations, they are still simple to understand and manipulate. In chapter 4.2., for generalized functional dependencies, the utilization of the solution of the implication problem is demonstrated for the axiomatization of functional, dual functional and monotone functional dependencies. Armstrong axioms are shown to be tautologies in dependency generalized functional logic.

Almost all technical and complexity issues in dependency theory can be better analyzed utilizing our approach. We demonstrate this claim as follows:

1. There are other types of dependencies that imply functional dependencies and behave exactly like functional dependencies with respect to different properties such as lossnessness. It is also shown that many different types of generalized functional dependencies that may seen to deny the existence of functional dependencies, are in fact embedded functional dependencies. These types of constraints are covered by the dependency propositional logic, and its calculus but not by

Armstrong formal system.

2. The controversy about mixed functional and multivalued dependencies can be easily understood from the generalized functional dependency perspective.

3. As it is mentioned in the introduction, our approach is more suitable to study several technical issues in the theory of the relational database.

As already remarked, generalized functional dependencies reflect a refinement of the functional dependency concept. For $U = \{A,B,C,D\}$ consider the following generalized functional dependency $H =$

$((\neg A \wedge \neg B \wedge \neg C) \vee (\neg A \wedge B \wedge C) \vee (A \wedge \neg B \wedge C) \vee (A \wedge B \wedge \neg C))$.

Using theorem 4.1.4. we get that from H follows $\{A,B,C\} \to U$,i.e. $\{A,B,C\}$ is a key for any relation r with $r \mid\mid== H$. Furthermore, we get that this functional dependency is the only which is implied by H. Nevertheless, we can construct a relation r on U such that r satisfies the functional dependency $\{A,B,C\} \to U$, but r does not obey H . An example is the following relation r

A	B	C	D
1	1	0	1
0	1	1	2
1	0	1	3

.

We can observe that constraints like H behave exactly like functional dependencies. Consider further a dependency set containing only the generalized functional dependency $A \to \neg B$ for $U = \{A,B,C\}$, i.e. each two different tuples t, t' which are equal on A should be different on B. Clearly, the constraint indicates that B is not functionally dependent on A . Thus, it may be thought that the initial set of functional dependencies is empty. Using theorem 4.1.4., it is not difficult to show that the given constraint implies that $\{A,B\}$ is a key for U . This constraint determines also that $\{A\}$ and $\{B\}$ are not keys. The richness of the language of generalized functional dependencies uncovers many interesting types of constraints. The study of the mathematical structure of these constraints is worth investigation. Additionally, these constraints may be utilized in certain issues such as horizontal decomposition of relations and query processing.

The introduced classes of dependencies can be and at present are used for an improvement of friendliness of user languages and of user design languages and design systems at present. Most of languages proposed only idiosyncratic versions of operations of the relational calculi. General functional dependencies and generalized functional dependencies can be therefore used for a more powerful and user-well-intentioned, nearly natural language design of databases. The variety of different dependency classes can be grouped into three main groups: 1. reality dependencies, i.e. dependencies which are used in reality for the database design, e.g. functional, inclusion, exclusion, multivalued dependencies; 2. database dependencies, i.e. dependencies which are used for the representation of the database, e.g. join, tuple-generating dependencies; 3. design dependencies, i.e. dependencies which can be used for a user-friendly schema design, e.g. general functional and generalized functional dependencies.

The importance of design dependencies can be explained and illustrated in the following contents. The classical theory neglects to distinguish between dependencies that reflect structural properties of the data and those that are merely integrity constraints. For instance, the functional dependency {A,B} -> {C} can be considered in different contexts:

1. It holds also {A} -> {B} and therefore {A} -> {C} .
2. It holds also {A} -> {B} , {B} -> {A}, and therefore {A} -> {C}, {B} -> {C} .
3. It is not valid that {A} -> {B}, and also {A} -> {C}, {B} -> {C} .
4. It is not valid that {A} -> {B}, {B} -> {C}, but it holds {A} -> {C}.
5. It is not valid that {A} -> {B}, {A} -> {C} , but it holds {B} -> {C}.
6. It is not valid that {A} -> {B}, {A} -> {C} , {B} -> {C} .

Our approach takes into consideration the different roles of functional dependencies. For instance, case 6 denotes the fact that there is no close relationship between {A}, {B} and {C} but only between {A,B} and {C}. Design dependencies must be powerful enough to represent these different meanings of functional dependencies. Another problem in scheme design is that the dependencies may represent not the presence or absence of relationships between the attributes, but rather constraints which have little influence on the way the data should be structured. This distinction is due to /BEKI 86/. The phenomenon there explained by the fact that a dependency as used in the classical design theory is intended to express both a basic relationship and an integrity constraint. Reality dependencies are primarily used to represent pure integrity constraints. Database dependencies are used for the representation of basic and indirect relationships which are significant in the scheme design.

4.2. PROPERTIES OF FUNCTIONAL DEPENDENCIES

In the first part of this section, we discussed generalized functional dependencies. Functional dependencies are special generalized functional dependencies.

Example 1.3. Despicts the relation cinema-information with
U = {CINEMA, ADDRESS, DATE, TIME, FILM}.
This relation tells in which cinema which film is shown. Not every combination of cinema, addresses, dates, times and films is to be found. The following restrictions apply, among others.

1. For each cinema, there is exactly one address.
2. For any given cinema, data and time, there is only one film.

These restrictions are examples of functional dependencies. Informally, a functional dependency occurs when the values of a tuple on one set of attributes uniquely determine the values on another set of attributes.

Our restrictions can be phrased as

{CINEMA} --> {ADDRESS}

{CINEMA,DATE,TIME} --> {FILM}.

A subset of the relation cinema-information is presented in the following table.

CINEMA	ADDRESS	DATE	TIME	FILM
Schauburg	Buchwitz-Str.	daily	18	Tootsie
Schauburg	Buchwitz-Str.	daily	21	Le Bal
Ost	Wehlener Str.	Mo-We	17	Mephisto
Ost	Wehlener Str.	Mo-We	20	A Chorus Line
Ost	Wehlener Str.	Th-Su	20	Stalker
Park	Bautzener Str.	daily	9	Alice
Park	Bautzener Str.	daily	18	Winnetou

The concepts and results of the second part of this section are either published (see for example /CODD 70/. /ARM 74/. /DEKA 83/) or belong to the folklore. There we use, a short approach of /DEKA 83/ applying methods of discrete mathematics.

Delobel and Casey /DECA 73/ gave a set of inference rules, which Armstrong /ARM 74/ showed were complete and correct. He also gave a method for constructing an Armstrong relation for a set of FD's (see also /DEGY 81/). The number of FD's that can be applied to a relation R is finite since there is only a finite number of subsets of U. Thus, it is always possible to find all the FD's that R satisfies, by trying all possibilities of pairs of elements of R. This approach is time-consuming. Certain dependencies of a relational database are known by its designer. We call these dependencies initial dependencies. In general, initial dependencies imply new dependencies. We now introduce a method to find the dependencies implied by a given set of initial functional dependencies.

We present now the formal system $\Gamma_{1,FD}$ /ARM 74/.

Axioms (FD0) $X\,Y \to Y$ for $X,Y \subseteq U$

Rules

(FD1)(transitivity) $$\frac{X \to Y\ ,\ Y \to Z}{X \to Z}$$ for $X,Y,Z \subseteq U$

(FD2)(augmentation) $$\frac{X \to Z}{X\,Y \to Y\,Z}$$ for $X,Y,Z \subseteq U$.

Theorem 4.2.1. The system $\Gamma_{1,FD}$ is sound and complete for implication of FD's.

Theorem 4.2.6, lemma 4.2.5, 4.2.7 and 4.2.8 prove theorem 4.2.1. Another proof of theorem 4.2.1 uses theorem 4.1.4 only.

From the rules of the formal system $\Gamma_{1,FD}$, it is easy to prove the soundness of following inference rules.

(FD3)(union) $$\frac{X \to Y\ ,\ X \to Z}{X \to YZ}$$ for $X,Y,Z \subseteq U$

(FD4)(projection) $$\frac{X \to Y}{X \to Z}$$ $X,Y,Z \subseteq U$, $Z \subseteq Y$

(FD5)(pseudotransitivity) $$\frac{X \to Y\ ,\ Y\,Z \to V}{X\,Z \to V}$$ for $X,Y,Z,V \subseteq U$.

There are also other sound and complete formal systems, for example $\Gamma_{2,FD}$. Formal system $\Gamma_{2,FD}$.

Axiom. (FD0') $X \to \{A\}$ for $X \subseteq U$, $A \in X$.

Rules (FD1)

(FD4).

The axiom (FD0') is a stronger version of (FD0). By theorem 4.2.1 and by theorem 4.1.4 holds

Corollary 4.2.2. 1) The system $\Gamma_{2,FD}$ is sound and complete for implication of

FD's.

2) For FD's $X \to Y$, $X' \to Y'$

$\{X \to Y\} \models X' \to Y'$ iff $Y' \subseteq X'$ or $X \subseteq X'$ and $Y' \subseteq Y\, X'$.

3) For FD's $X \to Y$, $X' \to Y'$ with $X \cap Y = X' \cap Y' = \emptyset$,

$\{X \to Y\} \models X' \to Y'$ iff $X \subseteq X'$ and $Y' \subseteq Y$.

4) If the FD $V \to W$ is derived from C using $X \to Y$ '

then $\models V \to X$.

Define the function L on U by

$L_r(X) = \{B \mid r \mid\mid== X \to \{B\}\}$

and for a set of functional dependencies C

$L_C(X) = \{B \mid C \models X \to \{B\}\}$.

These functions possesses some simple properties:

Lemma 4.2.3. Let $X, Y \subseteq U$. Then

(2.1) $X \subseteq L_r(X)$;

(2.2) $X \subseteq Y$ implies $L_r(X) \subseteq L_r(Y)$;

(2.3) $L_r(L_r(X)) = L_r(X)$.

(2.1') $X \subseteq L_C(X)$;

(2.2') $X \subseteq Y$ implies $L_C(X) \subseteq L_C(Y)$;

(2.3') $L_C(L_C(X)) = L_C(X)$.

Proof. (2.1) is obvious. It means that $X \to \{B\}$ holds for all $B \in X$. Indeed, if two tuples are equal in X, they must be equal in B, as well.

To prove (2.2), suppose that $A \in L_r(X)$, that is $r \mid\mid== X \to \{A\}$. In other words, any two tuples which are equal in X, coincide also in A. $X \subseteq Y$ implies that X can be replaced by Y in the latter statement, so $r \mid\mid== Y \to \{A\}$, that is, $A \in L_r(Y)$ as we wanted to show it.

The part $L_r(L_r(X)) \subseteq L_r(X)$ is a consequence of (2.1). We have to prove $L_R(L_R(X))$ $L_r(X)$, only. Let $A \in L_r(L_r(X))$. Then any two tuples in $L_r(X)$ are also equal in A. Consider now two tuples known to be equal in X. By definition, these two tuples must be equal in $L_r(X)$, therefore in A, i.e. $A \in L_r(X)$. The proof is complete.

To prove (2.1') - (2.3') is left to reader. For the proof can be used the results of chapter 3.1.

The literature of discrete mathematics calls a function satisfying (2.1)-

(2.3) a <u>closure</u>. Lemma 4.2.3 enables us to call L_r and L_C a closure.

Now we consider another relation between the closure and the dependencies. The next lemma can be easily proved.

<u>Lemma 4.2.4</u>. Let $X, Y \subseteq U$, r a relation on U.
$r \models X \rightarrow Y$ iff $Y \subseteq L_r(X)$.

Lemma 4.2.3 and 4.2.4 imply the following properties of the dependencies.

<u>Lemma 4.2.5</u>. Let $X, Y, Z \subseteq U$, r a relation on U.

(2.4) $r \models X \rightarrow X$;

(2.5) $r \models X \rightarrow Y$ and $r \models Y \rightarrow Z$ imply $r \models X \rightarrow Z$;

(2.6) $X \subseteq X'$, $Y' \subseteq Y$ and $r \models X \rightarrow Y$ imply $r \models X' \rightarrow Y'$;

(2.7) $r \models X \rightarrow Y$ and $r \models Z \rightarrow W$ imply $r \models XZ \rightarrow YW$.

Proof. (2.4) is a consequence of Lemma 4.2.4 and (2.1). By lemma 4.2.4 $r \models X \rightarrow Y$ can be written in the form $Y \subseteq L_r(X)$. (2.2) implies $L_r(Y) \subseteq L_r(L_r(X))$ and hence we have $L_r(Y) \subseteq L_r(X)$ because of (2.3). $r \models Y \rightarrow Z$ is equivalent to $Z \subseteq L_r(Y)$, therefore $Z \subseteq L_r(X)$ follows. This yields $r \models X \rightarrow Z$, again by lemma 4.2.4 (2.5) is proved.

Prove now (2.6) $X \rightarrow Y$ is equivalent to $Y \subseteq L_r(X)$. $Y' \subseteq Y$ implies $Y' \subseteq L_r(X)$. (2.2) and $X \subseteq X'$ result in $L_r(X) \subseteq L_r(X')$, and hence we have $Y' \subseteq L_r(X')$ which is equivalent to the wanted $r \models X' \rightarrow Y'$.

The condition of (2.7) can be rewritten into the forms $Y \subseteq L_r(X)$ and $W \subseteq L_r(Z)$. Hence, we obtain $YW \subseteq L_r(X)\, L_r(Z)$. (2.2) yields $L_r(X) \subseteq L_r(XZ)$ and $L_r(Z) \subseteq L_r(XZ)$ can be obtained similarly. These imply $YW \subseteq L_r(XZ)$ which is equivalent to $r \models XZ \rightarrow YW$.

Suppose now, in general, that a system of pairs (X,Y) of subsets of U is given which complies with the conditions (2.4)-(2.7). Such a system is called <u>full F-family</u>.

Lemma 4.2.5 points out the fact that dependencies form a full F-family.

In this way, we associated a full F-family with each relation. It is easy to see that the same full F-family can be associated with several different relations. On the other hand, as we see later, there is at least one relation to any full F-family.

Now we want to characterize full F-families.

F-characterization. Let F be a set of FD's. Then, we say that F satisfies the F-characterization if for any $X,Y \subseteq U$, $X \to Y \notin F$ there is a $Z \subseteq U$ such that

(i) $X \subseteq Z$ and $Y \not\subseteq Z$;

(ii) if $X' \to Y' \in F$ and $X' \subseteq Z$ then $Y' \subseteq Z$.

Now we can prove the following characterization theorem for full F-families.

Theorem 4.2.6. Let $F \subseteq Pow(U) x Pow(U)$. Then F satisfies the F-characterization iff F is a full F-family.

Proof. Suppose that F satisfies the F-characterization. Then:

(2.3) If $(X,X) \notin F$ then there is a $Z \subseteq U$ such that $X \subseteq Z$ and $X \not\subseteq Z$ which is a contradiction.

(2.4) If $(X,Y) \in F$, $(Y,Z) \in F$ and $(X,Z) \notin F$, then there is a $V \subseteq U$ such that $X \subseteq V$ and $Z \not\subseteq V$. Furthermore $(X,Y) \in F$, $X \subseteq V$ imply $Y \subseteq V$ and using $(Y,Z) \in F$, $Z \subseteq V$ which is a contradiction.

The proof of (2.5), (2.6) is analogous.

Suppose now that F is a full F-family. Let $(X,Y) \in F$, $X,Y \subseteq U$. Obviously, $(U,U) \in F$ by (2.3). Thus by (2.5) $(U,Y) \in F$ holds. $X \subseteq U$ and $(X,Y) \notin F$, consequently, there is an $Z \subseteq U$ which is maximal w.r.t. the property $(Z,Y) \notin F$ and $X \subseteq Z$. Let Z, $X \subseteq Z$, be a set such that $(Z,Y) \notin F$ and Z' with $Z \subsetneq Z'$ implies $(Z',Y) \in F$. We state now that Z satisfies (i) and (ii) of the F-characterization. That is, by the choice of Z, $X \subseteq Z$ holds. By (2.3) and (2.5) $Y \subseteq Z$ implies $(Z,Y) \in F$. Thus, we have $Y \not\subseteq Z$. Let $(V,W) \in F$ and $V \subseteq Z$.

$W \not\subseteq Z$ implies for $Z' = WZ$ $Z' \neq Z$ and by maximality of Z $(Z',Y) \in F$ holds. $(Z,Z) \in F$ by (2.4), hence (2.7) implies that $(Z,Z') \in F$. Now $(Z,Z') \in F$ and $(Z',Y) \in F$ and (2.5) imply that $(Z,Y) \in F$ which is a contradiction.

We can also prove a stronger characterization theorem for full F-systems. For that, following definition /THAL 83/, DEGY 81/ is required. Let $\underline{X} = \{X_1,\dots,X_m\}$ be a set system. Then $\underline{X}$ is a Φ-system, if for any i,j,k,l, $1 \leq i,j,k,l \leq m$, $i \neq j$, $k \neq l$, $X_i \cap X_j = X_k \cap X_l$.

Strong F-characterization. Let $F \subseteq Pow(U) x Pow(U)$. Then we say that F satisfies the strong F-characterization if there is a natural number k and an indexed set of subsets of U, $\{E_{ij} \mid 1 \le i < j \le k\}$ such that

(i') If $(X,Y) \notin F$, $X,Y \subseteq U$ then there are i,j such that $X \subseteq E_{ij}$ and $Y \not\subseteq E_{ij}$.

(ii') If $(X,Y) \in F$ and for some i,j $X \subseteq E_{ij}$ then $Y \subseteq E_{ij}$.

(iii') For any $1 \le i < j < l \le k$ $\{E_{ij}, E_{il}, E_{jl}\}$ is a Φ-system.

Lemma 4.2.7. If $F \subseteq Pow(U) \times Pow(U)$ satisfies the F-characterization then F satisfies the strong F-characterization.

Proof. Suppose, that F satisfies the F-characterization. For any $(X,Y) \notin F$, $X,Y \subseteq U$ take an $E(X,Y) \subseteq U$ guaranteed by the F-characterization. List these $E(X,Y)$'s as $E_2,\dots,E_k$. For $1<j\le k$ let $E_{1j} = E_j$ and for $1<i<j\le k$ let $E_{ij} = E_i \cap E_j$. Obviously, $\{E_{ij} \subseteq U \mid 1\le i<j\le k\}$ demonstrates that F satisfies the strong F-characterization.

Lemma 4.2.8. Let $F \subseteq Pow(U) \times Pow(U)$ satisfies the strong F-characterization. Then there is a relation r on U with $F = \{(X,Y) \mid X,Y \subseteq U,\ r \models X \dashrightarrow Y\}$.

Proof. Let $\{E_{ij} \mid 1\le i<j\le k\}$ show that F satisfies the strong F-characterization. We construct the tuples of r by induction.

Let $t_1(A) = 0$ for $A \in U$.

Suppose that $m < k$ and the tuples $t_1,\dots,t_m$ have been constructed so that for each $1\le i<\le m$ $E_{ij} = \{A \mid t_i(A) = t_j(A)\}$. Then

$$t_{m+1}(A) = \begin{cases} r_j(A) & \text{if } A \in E_{j(m+1)} \text{ for some } 1\le j\le m \\ m & \text{else} \end{cases} .$$

Now $A \in E_{i(m+1)} \cap E_{j(m+1)}$ implies $t_i(A) = t_j(A)$ because E_{ij}, $E_{i(m+1)}$ and $E_{j(m+1)}$ form Φ-systems and the induction hypothesis holds for $i,j \le m$.

If for $1\le i\le m$ $A \notin E_{i(m+1)}$ then $t_i(A) \ne t_{m+1}(A)$. Let $r = \{t_1,\dots,t_k\}$. The proof is complete.

It is useful, for database logical design, normalization and effective algorithms, to utilize the full information on given relations. It is well known that functional dependencies are the favorite constraints used to decompose relation schemes. This privilege is certainly due to the simplicity of the concept of functional dependencies, and to their wide-spread appearance in the real world. However, in a great number of applications there is a requirement to allow violation of some FD's, i.e. functional dependencies that are desired, but that do not hold in the relation.

The constraint

$\exists x \exists y \exists y' \exists z \exists z' (P(x,y,z) \wedge P(x,y',z') \quad y \neq y')$

is called <u>excluded functional constraint</u> (briefly EFD) and for

$X = \{A_i \in U \mid x_i \text{ in } x\}$, $Y = \{A_i \in U \mid y_i \text{ in } y\}$

denoted by $X \not\rightarrow Y$.

Obviously, for a relation $r \models X \not\rightarrow Y$ iff $r \not\models X \rightarrow Y$.

For a detailed examination of such systems, we can use the approach of /DEBR 85/, the concept of conflict free sets. In /THAL 84/ a formal system for FD's and excluded FD's is presented and proved its soundness and completeness.

Formal system $\Gamma_{FD,EFD}$

Axioms $X \rightarrow X$ for $X \subseteq U$.

Rules For subsets $X,Y,Z,W,V \subseteq U$

(FDEFD1) $\dfrac{X \rightarrow Y \ , \ Y \rightarrow Z}{XVW \rightarrow ZW}$

(FDEFD2) $\dfrac{X \rightarrow Y \ , \ XVW \not\rightarrow ZW}{Y \not\rightarrow Z}$

(FDEFD3) $\dfrac{Y \rightarrow Z \ , \ XVW \not\rightarrow ZW}{X \not\rightarrow Y}$ $\quad Z \neq \emptyset$.

For horizontal decomposition (chapter 8), so-called afunctional and antifunctional dependencies are introduced in /DBRA 85/.

Let X be a set of attributes.

A set of tuples r' in a relation r is called <u>X-complete</u> iff

$r'[X] \cap (r-r')[X] = \emptyset$.

Let X,Y,Z be sets of attributes. $X,Y,Z \subseteq U$.

The <u>antifunctional dependency</u> $X \not\not\rightarrow^{Z} Y$ means that in every non-empty Z-complete set of tuples in a relation r the functional dependency $X \rightarrow Y$ does not hold.

Clearly, it holds

$X \not\not\to^{U} Y \models X \not\to Y$ and $X \not\to Y \not\models X \not\not\to^{U} Y$.
Defining r as the $\emptyset$-complete set the excluded FD $X \not\to Y$ can be represented as a special antifunctional dependency $X \not\not\to^{\emptyset} Y$.

The antifunctional dependency $X \not\not\to^{x} Y$ is also called <u>afunctional dependency</u> and denoted by $X \not\not\to Y$. This dependency is equivalent to the following formula for corresponding sequences of variables

$$\forall x \exists y \exists y' \exists z \exists z' (P(x,y,z) \wedge P(x,y',z') \wedge y \neq y').$$

It is of interest that a sound and complete formal system exists for sets of functional and afunctional dependencies which is analogous to $\Gamma_{FD,EFD}$.

Now we want to give a combinatorial characterization of the sets which are of minimal cardinality with respect to the property that they imply all the dependencies of a given full F-family.

By this problem it is tried to determine the most "complex" system of depdencies in a database with n attributes. Due to the presented results we can speak about full F-families instead FD's.

Let F be a full F-family. The dependency $X \to Y$ F is called <u>basic</u> if

1) $X \neq Y$;
2) there are no $X' \subset X$, Y', $Y \subset Y'$, with $(X',Y) \in F$ or $(X,Y') \in F$.

All FD's trivially follows from the basic dependencies. Therefore, their number can be considered the complexity or the <u>design complexity</u> of the database. Thus, our aim to this part is in fact equivalent to the problem of finding the most complex database.(see also /BDHF 80/)

Let N(n) denote the maximum number of basic dependencies in a database with n attributes.

It is easy to construct a relation in which the basic dependencies are of the form $X \to X\{A\}$ where A is a fixed attribute. That is, $2^{n-1} \leq N(n)$.

Now we show an upper estimate on N(n). Introducing the notation
$\tilde{F} = \{ X \mid (X,Y)$ is a basic pair in $F\}$,
let (X,Y) be a basic pair, and suppose that $X \subseteq Z \subseteq Y$, $|Z| = |X| + 1$. It is easy to see that $Z \notin \tilde{F}$. Such a Z can be obtained from at most n different sets X, consequently for at least $|\tilde{F}|/n$ sets Z holds $Z \notin \tilde{F}$. This implies $|\tilde{F}| + |\tilde{F}|/n \leq 2^{n}$. Hence we have

Corollary 4.2.9. $2^{n-1} \leq N(n) \leq 2^n (1 - 1/(n+1))$.

In /DEKA 83/ a stronger result is proved using /KOST 84/.

$$2^n (1 -(\log_2\log_2(n))/(\log_2(e) \log_2(n)))(1+o(1)) \leq N(n) \leq$$
$$2^n(1 - (\log_2(n))^{3/2}/(150\, n)).$$

One question remains unsolved; what are better bounds of N(n) ?

Finally we give the combinatorial characterization of sets which are of minimal cardinality w.r.t. the property that they imply all the dependencies of a given full F-family.

Let F be a full F-family. A subset F' of F is called minimal generating subset of F if $F = \{X \dashrightarrow Y \mid F' \models X \dashrightarrow Y\}$ and if there is no subset F" of F' which is a minimal generating subset of F.

All dependencies of F follow from some minimal generating subset F'. Therefore the size of F' can be considered the design complexity of the database.

Thus, our aim of this part is now in fact equivalent to the problem of finding the most complex full families F.

Let $N^*(F)$ denote the minimal size of a minimal generating subset of F and let $N^*(n)$ denote the maximum size of $N^*(F)$ for full F-families F in a database with n attributes.

Example 4.2.10. Let

$$C = \{X \dashrightarrow \{A_n\} \mid |X| = [(n-1)/2],\ X \subseteq \{A_1,\ldots,A_{n-1}\}\}$$

([t] denotes the integer part of t).

Then C is a minimal generating subset of

$$C^+ = \{XY \dashrightarrow Y\{A_n\} \mid |X| \geq [(n-1)/2],\ X \subseteq \{A_1,\ldots,A_{n-1}\},\ Y \subseteq U\}$$
$$\cup \{ XY \dashrightarrow Y \mid X,Y \subseteq U \} .$$

We get the lower estimate on $N^*(n)$

$$\binom{n}{\lfloor \frac{n-1}{2} \rfloor} \leq N^*(n) .$$

Lemma 4.2.11. If $\{X_1 \dashrightarrow Y_1, X_2 \dashrightarrow Y_2,\ldots,X_m \dashrightarrow Y_m\} \models X \dashrightarrow Y$ then there is a number i with $X_i \subseteq X$.

Proof. For the proof we use the system $\Gamma_{1,FD}$ and theorem 4.2.1.

Assume that $\{X_1 \dashrightarrow Y_1, X_2 \dashrightarrow Y_2, \ldots, X_m \dashrightarrow Y_m\} \vdash X \dashrightarrow Y$ holds. It is easy to deduce by mathematical induction on derivation degree the property of the lemma. For derivation degree 0, it is obvious. If the property of the lemma is proved for derivation degree k then we get all new dependencies in the next derivation step by using the axiom or the rules (FD1) or (FD2). Therefore, the lemma holds for derivations of derivation degree k+1.

Directly by lemma 4.2.11 and corollary 4.2.9 we obtain

$$2^{n-1}/\sqrt{n} \leq N^*(n) \leq 2^n (1 - 1/(n+1)) .$$

It is easy to prove that for any full F-family C there exists a minimal generating subset C' of C such that C' is a set of basic dependencies. Using the following example and the inequality

$$\binom{n}{\lfloor \frac{n}{2} \rfloor} \geq \frac{2^{n-1}}{\sqrt{n}}$$ we get the lower bound.

Example 4.2.12. Let $F = \{X \dashrightarrow Y \mid X,Y \subseteq U, X \cap Y = \emptyset, |X| = \lfloor n/2 \rfloor\}$. Then F is a minimal generating subset of
$F^+ = \{X \dashrightarrow Y \mid X,Y \subseteq U, |X| \geq \lfloor n/2 \rfloor\} \cup \{X \dashrightarrow Y \mid Y \subseteq X, X \subseteq U\}$.
Because by lemma 4.2.11 we can prove that $F' \models X \dashrightarrow Y$ for any dependency $X \dashrightarrow Y$ of $F' = F^+ - F$.

Using theorem 4.1.6 we obtain now

Corollary 4.2.13. $N^*(n) = N(n)$.

Using example 4.2.10 we get that there is a different size of minimal generating subsets of a given class.

Let $N_*(F)$ denote the maximal size of a minimal generating subset of F and

$$N^\wedge(F) = \frac{N_*(F)}{N^*(F)} , \quad N^\wedge(n) = \max_F N^\wedge(F) .$$

$N^\wedge(n)$ is called the dispersion of the class of FD's.
Using $F = \{\{A_1\} \dashrightarrow X \mid X \subseteq U\}$ we obtain the trivial

Corollary 4.2.14. $N^\wedge(n) \geq n-1$.

In /GOTT 87/ it is proved that $N^{\hat{}}(n) = n-1$.

4.3. HUNGARIAN AND MONOTONE FUNCTIONAL DEPENDENCIES

In /CZED 81/ and /DEGY 81/ generalizations of functional dependencies are introduced. In order to expound why we dealt with these concepts let us consider the following relation.

Example. Let U = {AUTHOR, TITLE, HALL, SHELF}. There is a library with eighteen books, three halls for different users and shelves in every hall. Given the following table.

AUTHOR	TITLE	HALL	SHELF
1	1	1	2
2	2	1	3
3	3	1	1
4	4	1	2
5	5	2	3
6	6	2	1
7	7	2	2
8	8	2	3
9	9	3	1
10	10	3	2
11	11	3	3
12	12	3	1
1	4	1	1
5	8	3	3
4	1	1	3
7	10	3	2
6	10	2	2
6	9	2	1

Thus, {AUTHOR, TITLE} $\rightarrow^{D}$ {HALL, SHELF} holds in r .

Now in connection with this example, we try to express why the concepts of dual, strong, weak and monotone functional dependencies can be of some practical importance.

The final purpose of any database system is to provide the user with actual information.

In any time-varying data structure at a particular moment of time there are dependencies. Some of them may be fortuitous or unimportant, but it is reasonable to require that at least certain dependencies should be present at any time. Organizing the data structure and some of the user's activities can be based on these initial dependencies. In case of functional dependencies these has been shown in Codd's papers /CODD 70/, /CODD 71/.

Now the following reasons have been collected to show the advantage of using more types of dependencies besides the functional or generalized functional one.

(1) The semantics of relations and databases can be given in a feeble form. There can be other types of generalized functional dependencies between attributes even if there is no functional one between them. The user can happen to know only at least one but not all the values of attributes in the "life". Just think of the visitor of the library in our example 4.3.1. If, for example, U is a set of several attributes of a criminal, say U = {length, age, citizenship,...} and r is a relation of a criminal data bank then a detective also can be such a user at the beginning of his investigation.

Sometimes, the user can require only the value of some attributes and the relationship between these attributes.

(2) More powerful dependencies are more useful for database design.

Sometimes the information supply can be accelerated by describing a particular dependency with coding functions or functions. The only requirement tailored to those functions is that they should be computed easily or stored in relatively small tables. For instance, in example 4.3.1, the dependency {AUTHOR, TITLE} --> {HALL, SHELF} is described by the functions $[(i-3)/4]$ and $1 + 3\{i/3\}$ ($\{x\}$ debits the fraction part of x). The functional dependency {AUTHOR, TITLE} --> {HALL, SHELF} also holds in our example. Consequently, there exists a function which describes this dependency.

But the table of this function is the table of r itself, and so scanning the whole table cannot be avoided in this way. I.e., sometimes it is not the functional dependency which yields the most economic way of information supply.

As mentioned in /STPA 84/, Hungarian functional dependencies can be used also for access authorization, for data maintenance, for query optimization based on generalized functional dependencies, and for efficient verification of integrity constraints.

(3) Generalized functional dependencies are useful for describing upper and lower bounds of existence of functional dependencies. Strong functional dependencies are systems of functional dependencies with small left sides. Dual functional dependencies are negative restrictions for key dependencies (keys). Weak functional dependencies are negative restrictions for functional dependencies. Monotone functional dependencies describe systems of weak, dual, strong and functional dependencies.

In order to investigate the various dependencies the first step is the axiomatization of families of such dependencies.

Using theorem 4.1.4. the known axiomatizations of different classes of special functional dependencies can be derived. We illustrate this application for dual functional dependencies. Dual functional dependencies are general functional dependencies, therefore only rules of the form $\beta_1, \beta_2 \vdash \beta_3$ and $\beta_1 \vdash \beta_2$ are needed. From the theory of Boolean functions there is known that x_1 , $x_1 \vee x_2$ forms a complete set of disjunctions. Therefore, only corollary 5 and the consideration of dependencies $X Z \rightarrow^D Y V$, $X Z \rightarrow^D Y \cap V$, $X \cap Z \rightarrow^D Y V$, $X \cap Z \rightarrow^D Y \cap V$ is required.

Another proof can be found applying the approach of section 4.2 /DEGY 81/. Therefore, some proof of the following theorems can be omitted.

We present now the formal system $\Gamma_{1,DFD}$.

Axioms: $X \rightarrow^D X Y$ for $X,Y \subseteq U$;

Rules: For $X, Y, Z \subseteq U$

(DFD1) $$\frac{X \dashrightarrow^D Y \ , \ Y \dashrightarrow^D Z}{X \dashrightarrow^D Z}$$ (transitivity)

(DFD2) $$\frac{XY \dashrightarrow^D Z}{X \dashrightarrow^D Z}$$ (augmentation)

(DFD3) $$\frac{X \dashrightarrow^D Z \ , \ Y \dashrightarrow^D Z}{X Y \dashrightarrow^D Z}$$ (union)

(DFD4) If $X \dashrightarrow^D \emptyset$ then $X = \emptyset$ (metarule)

No other combinatorial combinations which are not implied by this set can be used for valid implications. Therefore this set forms a complete set.

From the presented rules of the system $\Gamma_{1,DFD}$ it is easy to prove the soundness of other inference rules, for instance

(DFD5) $$\frac{XY \dashrightarrow^D Z}{X \dashrightarrow^D ZV}$$ (full augmentation)

(DFD6) $$\frac{X \dashrightarrow^D Y \ , \ Z \dashrightarrow^D V}{XZ \dashrightarrow^D VY}$$ (full union)

(DFD7) $$\frac{V \dashrightarrow^D XZ \ , \ X \dashrightarrow^D Y}{V \dashrightarrow^D YZ}$$ (pseudotransitivity)

<u>Theorem 4.3.1</u>. The system $\Gamma_{1,DFD}$ is sound and complete for implication of

DFD's.

The proof is analogous to proof of theorem 4.2.1.
We use the

<u>D-characterization</u>. Let F be a set of dual functional dependencies. Then we say that F satisfies the D-characterization if for any $X,Y \subseteq U$ with $X \dashrightarrow Y \notin F$ there is a $Z \subseteq U$ such that

(i) $X \cap Z \neq \emptyset$, $Y \cap Z = \emptyset$;

(ii) if $X' \dashrightarrow^{D} Y' \in F$ and $X' \cap Z \neq \emptyset$ then $Y' \cap Z \neq \emptyset$.

We present now sound and complete formal systems Γ_{SFD}, Γ_{WFD}, Γ_{MFD} for strong functional dependencies, weak functional dependencies and monotone functional dependencies. The proofs are analogous to the proof of theorem 4.2.1.
For these dependencies dependencies of the form $\emptyset \dashrightarrow^{H} Y$ for $H \in \{S,W,M\}$ should be also considered especially since they mean that any two tuples of a relation agree under H . Dependencies of the form $X \dashrightarrow^{H} \emptyset$ are trivial.

Formal system Γ_{SFD} .
Axiom $\{A\} \dashrightarrow^{S} \{A\}$ for $A \in U$;
Rules. For $X,Y,Z,V,W \subseteq U$

(SFD1) $\dfrac{X \dashrightarrow^{S} Y \ , \ Y \dashrightarrow^{S} Z}{X \dashrightarrow^{S} Z}$ (transitivity)

(SFD2) $\dfrac{XV \dashrightarrow^{S} YW}{X \dashrightarrow^{S} Y}$ $X \neq \emptyset$ (augmentation)

(SFD3) $\dfrac{X \dashrightarrow^{S} Y \ , \ V \dashrightarrow^{S} W}{X \cap V \dashrightarrow^{S} YW}$ $X \cap V \neq \emptyset$ (intersection-union)

(SFD4) $\dfrac{X \dashrightarrow^{S} Y \ , \ V \dashrightarrow^{S} W}{XV \dashrightarrow^{S} Y \cap W}$ (union-intersection)

For weak functional dependencies, it is easy to improve the known formal systems /DEGY 81/ using theorem 4.1.6.

A family of weak functional dependencies C is called (X,Y)-upright if there is a set Z with $X \subseteq Z$, $Z \cap Y = \emptyset$ and $C = \{X' \dashrightarrow^{W} U-X' \mid X \subseteq X' \subseteq Z\}$.

Formal system Γ_{WFD} .

Axiom $X \dashrightarrow X$ for $X \subseteq U$.

Rules. For $X,Y,V,W \subseteq U$

(WFD1) $\dfrac{C}{X \dashrightarrow^{w} Y}$ if C is (X,Y)-upright (upright rule)

(WFD2) $\dfrac{X \dashrightarrow^{w} Y}{XV \dashrightarrow^{w} YW}$ (augmentation).

The weak functional dependencies are influential functional dependencies. The following corollary characterizes families of functional dependencies. This corollary follows easily from theorem 4.1.6.

<u>Corollary 4.3.2</u>. Given a system C of GD's with

$C \not\models x_1 \wedge x_2 \wedge \ldots \wedge x_n \dashrightarrow \emptyset$, $C \not\models 1 \dashrightarrow x_1 \vee \ldots \vee x_n$

(i.e. $C \not\models U \dashrightarrow \emptyset$, and $C \not\models \emptyset \dashrightarrow U$). Then there exists an equivalent to system C' of weak functional dependencies.

In /KLIP 83/ a sound and complete formal system for monotone functional dependencies is given. The soundness and completeness of the formal system Γ_{MFD} follows easily from theorem 4.1.4.

An equivalent consideration can be used to prove the completeness and soundness of the following formal system for monotone functional dependencies. Let Pow(U) denote the set of all subsets of U and $Pow^{+}(U)$ the set of all non-empty subsets of U .

Given sets $\underline{X}$, $\underline{Y} \subseteq Pow(U)$, let $\underline{X} \cdot \underline{Y}$ denote the set

$\{XY \mid X \in \underline{X}, Y \in \underline{Y}\}$.

Then we get

Formal system Γ_{MFD} .

Axiom $\underline{X} \dashrightarrow^{M} \underline{XY}$ for $\underline{X} \subseteq Pow^{+}(U)$, $\underline{Y} \subseteq Pow(U)$;

Rules. For $\underline{X}, \underline{Y}, \underline{Z} \subseteq Pow^{+}(U)$, $\underline{V} \subseteq Pow(U)$

(MFD1) $\dfrac{\underline{X} \dashrightarrow^{M} \underline{Y} \, , \, \underline{Y} \dashrightarrow^{M} \underline{Z}}{\underline{X} \dashrightarrow^{M} \underline{Z}}$ (transitivity)

(MFD2) $\dfrac{\underline{X} + \underline{V} \dashrightarrow^{M} \underline{Z}}{\underline{X} \dashrightarrow^{M} \underline{Z}}$ (augmentation)

$$\frac{\underline{X} \to^{M} \underline{Y},\ \underline{Z} \to^{M} \underline{Y}}{\underline{X} u \underline{Z} \to^{M} \underline{Y}} \qquad \text{(union)} \tag{MFD3}$$

$$\frac{\underline{X} \to^{M} \underline{Y},\ \underline{X} \to^{M} \underline{Z}}{\underline{X} \to^{M} \underline{Y}\cdot\underline{Z}} \qquad \text{(product)} . \tag{MFD4}$$

4.4. KEY DEPENDENCIES

In databases, the keys play an important role. One of the suggestions for the handling of relations is the identification of sets of domains, called keys, which uniquely determine the values of remaining domains. In databases, the keys play an important role. The records or tuples can be uniquely found by them. A key is generally an attribute (or a combination of several attributes) which uniquely identifies a particular record without ambiguity. Of course, it is worth-while to consider the minimal ones, only. It is quite naturally to ask how many minimal keys exist in different relations. Delobel and Casey, Fadous and Forsyth, Ho Thuan, Luccesi and Osborn have given different algorithms for finding the set of all keys in relational databases given by a set of functional dependencies on the database. For characterizing the complexity of these algorithms we need some combinatorial bounds about the number of keys. We summarize some of the important combinatorial problems in relational databases, prove that the result of Demetrovics /DEME 79/ about the maximal number of minimal keys does not hold for finite domains and consider the maximal number of minimal keys about weighted domains. For practical purpose, keys are of different meaning and complexity. Domains for attributes have very different complexity. This is well known in practice but in theory of minimal keys, it is not taken into consideration. We prove that the maximal number of minimal keys in databases on nonuniform domains is also precisely exponential in the number of attributes but different in order from the maximal number of minimal keys on uniform domains.

At first, we consider the axiomatization of systems of keys. Remember that X is a <u>key</u> of a set C of FD's if it meets the following condition:
$C \vdash X \to U$. A key X is called <u>minimal key</u> for C if there is no proper subset X' of X with $C \vdash X' \to U$.

Now we present the following trivial system Γ_{KD} for key dependencies.

Formal system Γ_{KD} .

Axiom U --> U .

Rule $\frac{X \dashrightarrow U}{XY \dashrightarrow U}$ for $X,Y \subseteq U$ (augmentation) .

(KD1)

As an immediate consequence of theorem 4.1.4., we have the

Corollary 4.4.1. The system Γ_{KD} is sound and complete for implication of key dependencies.

Remember that by [m] is denoted the integer part of m .

Theorem 4.4.2./DEME 78/ The maximal number of minimal keys in a database with n attributes is $\binom{n}{[\frac{n}{2}]}$.

A set E of subsets of U is called Sperner system if for different elements X , Y of E the property $X \not\subseteq Y$ is valid.

Proof. The minimal keys K are subsets of U and do not include each other. The set of minimal keys forms a so-called Sperner family. Sperner's well-known theorem /SPER 28/ states that such a family can not contain more than $\binom{n}{[\frac{n}{2}]}$ members.

We will now construct an m-element relation r (with $m = \binom{n}{[\frac{n}{2}]-1} + 1$) having $\binom{n}{[\frac{n}{2}]}$ minimal keys.

The first tuple of r consists of nothing but 1's. The other tuples contain $[\frac{n}{2}] - 1$ 1's in all possible ways while the remaining entries of the i-th tuple are i's ($2 \leq i \leq \binom{n}{[\frac{n}{2}]-1} + 1$) . If we choose $[\frac{n}{2}]$ attributes in a tuple we find there only 1's or at least one number i different from 1. Therefore, the tuple i is uniquely determined. Any X with $X \subseteq U$, $|X| = [\frac{n}{2}]$ is a key. On the other hand, it is easy to see that no set X , $X \subseteq U$, with $|X| < [\frac{n}{2}]$ can be a key, the first tuple coincides with another one in r[X] . The proof is complete.

Example 4.4.3. The construction of the proof can easily be understood. For n = 4, see the relation r below:

A_1	A_2	A_3	A_4
1	1	1	1
1	2	2	2
3	1	3	3
4	4	1	4
5	5	5	1

Another relation with $\binom{4}{2}$ keys is the following:

A_1	A_2	A_3	A_4
1	1	1	1
1	2	2	2
2	1	2	3
3	2	1	3

It is easy to see that for n = 4 no relation r with only 3 tuples and $\binom{4}{2}$ minimal keys exists. Obviously, for n= 4 and the domain $\underline{D} = \{\{1,2\}\}$ there is no relation with $\binom{4}{2}$ minimal keys.

It is possible to give a more precise characterization of key systems for given sets of FD's /DETH 88/ (see also /HTLB 84/).

Let $C = \{ X_i \rightarrow Y_i \mid 1 \leq i \leq m\}$ be an FD system. Assume that C is reduced, i.e. $X_i \cap Y_i = \emptyset$, $1 \leq i \leq m$.

Let us denote

$X_C = X_1 X_2 \ldots X_m$; $Y_C = Y_1 Y_2 \ldots Y_m$;

$\underline{K}(U,C) = \{ X \subseteq U \mid C \vdash X \rightarrow U , X \text{ minimal key}\}$;

$X^+ = \{A \in U \mid C \vdash X \rightarrow \{A\}\}$ for $X \subseteq U$.

As an immediate consequence of definitions and theorem 4.1.4 we have the following

<u>Corollary 4.4.4</u>. Let $C = \{ X_i \rightarrow Y_i \mid 1 \leq i \leq m\}$ be a reduced FD system.

1. If $A \notin X_C$, and $C \vdash X \rightarrow Y$ then $C \vdash X-\{A\} \rightarrow Y-\{A\}$.
2. If $A \notin X$, $X \subseteq U$ and $C \vdash X \rightarrow \{A\}$ then $X\{A\}$ is not a minimal key.
3. If X is a minimal key then $U-Y_C \subseteq X \subseteq (U-Y_C)(X_C \cap Y_C)$.
4. $|U-Y_C| \leq |X| \leq |U-Y_C| + |X_C \cap Y_C|$.
5. If $Y_C - X_C \neq \emptyset$ then a nontrivial minimal key exists.
6. If $Y_C \cap X_C = \emptyset$ then $|\underline{K}(U,C)| = 1$ and $U-Y_C$ is the unique minimal key of C.
7. /FERN 84/ For any different $i,j \in \{1,2,\ldots,m\}$ $X_i ((U-X_i^+) \cap (X_j(U-X_j^+)))$ is a key of C.
8. /FERN 84/ The family $\{ X_i ((U-X_i^+) \cap (X_j(U-X_j^+))) \mid 1 \leq i,j \leq m , i \neq j\}$ can be used to find all minimal keys of C .
9. $\bigcap_{K \in \underline{K}(U,C)} K = U-Y_C$.
10. If $X_C \cap Y_C \neq \emptyset$ then $(U-Y_C)(X_C \cap Y_C)$ is not a minimal key of C .

Proof. 1., 2., 4., 5., 6., and 9. are obvious.

3. If X is a minimal key then obviously $X^+ = U$ and there fore $X^+ \subseteq XY_C$. This implies $U-Y_C \subseteq X$. Because it holds $U = (U-Y_C)(X_C \cap Y_C)(Y_C-X_C)$ it is sufficient

to prove that $X \cap (Y_C - X_C) = \emptyset$. If there exists an attribute $A \in X \cap (Y_C - X_C)$ then we get by 1. $C \vdash X - \{A\} \rightarrow U - \{A\}$, by (FDO) $C \vdash U - \{A\} \rightarrow X_C$ and by 2. $C \vdash X - \{A\} \rightarrow \{A\}$. By virtue of 2. X is not a minimal key. Therefore $X \subseteq (U - Y_C)(X_C \cap Y_C)$.

7. Let be i a fixed number. If $U - X_i^+ = \emptyset$ then we get $X_i = X_i ((U - X_i^+) \cap (X_j (U - X_j^+)))$ is a key of C .
If $U - X_i^+ \neq \emptyset$ then $((U - X_i^+) \cap (X_j (U - X_j^+))) \neq \emptyset$ for any j , $i \neq j$.
Now for j it is evident that
$C \models X_i ((U - X_i^+) \cap (X_j (U - X_j^+))) \rightarrow X_i^+ ((U - X_i^+) \cap X_j) ((U - X_j^+) \cap (U - X_i^+)))$
and consequently $C \models X_i ((U - X_i^+) \cap (X_j (U - X_j^+))) \rightarrow X_j (U - X_j^+)$.

8. It is easy to show that $K \subseteq X_i ((U - X_i^+) \cap (X_j (U - X_j^+)))$ for some i,j, $K \in \underline{K}(U,C)$ with $X_i \subseteq K$. We get the assertion using 7.

10. It is easy to see that by 3. and 9. the 10. is obvious.

Corollary 4.4.4 (especially 8.) can be used to design an interesting algorithm to find all keys for FD sets /FERN 84/.

<u>Example 4.4.5</u>. $U = \{A,B,H,G,Q,M,N,V,W\}$,
$C = \{\{A\} \rightarrow \{B\}, \{B\} \rightarrow \{H\}, \{G\} \rightarrow \{Q\}, \{V\} \rightarrow \{W\}, \{W\} \rightarrow \{V\}\}$. We get now
$X_C = \{A,B,G,V,W\}$, $Y_C = \{B,H,Q,V,W\}$, $X_C \cap Y_C = \{B,V,W\}$; $X_C - Y_C = \{A,G\}$,
$(X_C - Y_C)^+ = \{A,B,G,H,Q\}$, $U - Y_C = \{A,G,M,N\}$, $(X_C - Y_C)^+ \cap (X_C - Y_C) = \{A\} \neq \emptyset$;
$\underline{K}(U,C) \subseteq \{ X \mid \{A,G,M,N\} \subseteq X \subseteq \{A,G,M,N,V,W\}\}$ and using the Sperner-property we get $| \underline{K}(U,C) | \leq 2$. Using the algorithm implied by 8 of corollary 4.4.4 we get $\underline{K}(U,C) = \{\{A,G,M,N,V\},\{A,G,M,N,W\}\}$.

For Sperner systems and sets $\underline{K}$ of minimal keys, the set $\underline{K}^{-1}$ of <u>antikeys</u> /DETH 88/ can be defined as follows
$\underline{K}^{-1} = \{ X \subseteq U \mid \forall Y \in \underline{K} : Y \not\subseteq X$ and $\forall X' (X \subset X') \exists Y \in \underline{K} : Y \subseteq X' \}$.
It is easy to see that $\underline{K}^{-1}$ is also a Sperner system. Clearly, the elements of $\underline{K}^{-1}$ do not contain the elements of $\underline{K}$ and they are maximal for this property.

Let for $r = \{t_1, \ldots t_m\}$ $E_r = \{E_{ij} \mid 1 \leq i < j \leq m,\ E_{ij} = \{A \in U \mid t_i(A) = t_j(A)\}$. The set E_r is called equality system. Let be E' the maximal subset of E_r with the following property: if $X \in E'$ and $Y \in E_r$ then $X \not\subseteq Y$, i.e. the set of all maximal elements of E_r. The set E' is called maximal equality system of r .
Now we can prove the following theorem /DEGY 81/ , /DETH 88/.

Theorem 4.4.6. Let $\underline{K} = \underline{K}(U,C)$ be a non-empty Sperner-system and r be a relation on RS. Then $\underline{K}$ is the set of all minimal keys of r iff $\underline{K}^{-1}$ is the maximal equality system of r.

Proof. As $\underline{K}$ is a non-empty Sperner system, $\underline{K}^{-1}$ exists. $\underline{K}$ and $\underline{K}^{-1}$ are uniquely determined by each other.

1. Let $\underline{K}$ be the set of minimal keys of r, E' the maximal equality system of r. Since for any $Y \in \underline{K}$ and for any proper subset Y' of Y there exist two different tuples t, t' in r with $t[Y'] = t'[Y']$. Therefore, $Y'' \in E_r$ for Y'' with $Y' \subseteq Y'' \subset Y$. Furthermore, there exist a maximal Y'' with this property. According to the maximality of Y'' we get the following property:
If Y''' contains proper Y'' then for all different tuples t, t' of r $t[Y'''] \neq t'[Y''']$. Therefore Y''' is a key and $Y'' \in \underline{K}^{-1}$.

2. Assume that E' is the maximal equality system of r, i.e. for any key X of r $X \notin E'$ $(\forall\, t,t' \in r\colon t[X] \neq t'[X])$.
Let $\underline{K}$ be the set of all minimal keys of r. Let $X \in E'$. Then according to the definition of the set X, X is not a key of r. By definition of E' all Y containing proper X are keys. Consequently, by the definition of antikeys $X \in \underline{K}^{-1}$.
Let $X \in \underline{K}^{-1}$. Then there are different tuples in r with $t[X] = t'[X]$. According to the definition, X is maximal and $X \in E_r$. Therefore $X \in E'$.

We shall consider the number of minimal keys in restricted cases. In practical cases the domain is bounded. Therefore we need an upper bound for the maximal number of minimal keys in domain bounded databases.
A database r is called k-valued if no domain set in $\underline{D}$ contains more than k elements.
Let us denote by Fak(n) the number

$$\binom{n}{\lfloor \frac{n}{2} \rfloor} .$$

Theorem 4.4.7. The maximal number of minimal keys in k-valued databases is less then Fak(n) if $k^4 < 2n + 1$.

Proof. We shall prove that a relation r with Fak(n) minimal keys of size $m =$

$n/2$ does not exist for any natural n. By key properties and definitions it follows that a subset X of U exists with $|X| = m-1$ and with $t(X) = t'(X)$ for different elements t,t' of r. Since the Hamming-distance
$dis(t,t') = |\{ A \in U \ / \ t(A) \neq t'(A) \}|$ of different elements of r is not smaller than $d = n-m+1$, the relation r has not more than $M(n,d,k)$ elements, where $M(n,d,k)$ is the cardinality of maximal codes with distance d and elements from $\{1,2,\ldots,k\}^n$.
There is a well known bound /MWIS 77/ for $M(n,d,k)$:
$M(n,d,k) < k^n / n^t$ with $t \geq (d+1)/2$ and $t \leq (d+2)/2$.
For any subset X of U with $|X| = m-1$ there exists two elements t_X, t'_X in r with $t_X(X) = t'_X(X)$. All pairs (t_X, t'_X), (t_Y, t'_Y) are different for different sets X, Y. Otherwise we deduce a contradiction for $Z = X \ Y$. Now, we conclude that there exist at least $\binom{n}{m-1}$ different pairs of elements in r, i.e. $\binom{p}{2} \geq \binom{n}{m-1}$.
Define $f(k,n) = \frac{1}{2} k^{2n} / n^{m}$. From $p < k^n/n^t$ follows
$k^n:2n^t \, (k^n:n^t - 1) < f(k,n)$.
For $n = 2s + 1$ and $k^4 < 2n + 1$ we get $f(k,n) < \binom{n}{m-1}$
by $\binom{n}{k} > (\frac{n}{k})^n$
and for $n = 2s$ and $k^4 < 2n + 4$ we get
$f(k,n) < \binom{n}{m-1}$
by $\binom{n}{k} > (2 (n-k):(k+1))^k$.
That is a contradiction.

We remark that theorem 4.4.7 can be improved using this proof /THAL 84/.

Corollary 4.4.8. In k-valued databases with n attributes there are not more than $Fak(n) - n/2$ minimal keys for k with $k^4 < 2n + 1$.

We observed the equivalence between Sperner families and sets of minimal keys. This equivalence can be used for consideration of Armstrong relations.

There are also known some estimations on the average of keys in m-valued relations and the number of keys in almost all m-valued relations (see, for example, /SOLO 78/).

For practical purposes, keys with a low complexity are of special interest. In database literature there are known only few papers considering this important aspect in relational databases. Therefore, we need a complexity measure for the set U of attributes. But if a relation has different keys one of them can be distinguished as the most convenient. This can for instance be the shortest or more generally the key with the lowest complexity.

Example 4.4.9. Consider a student file. For each student, the department of student affairs is interested in the identity number, the name, the address, the attended courses with the corresponding marks and numbers in those courses (each student has his own number in each course), and the average grades. We can represent this information in a table called student.

IDNUMB	NAME	ADDRESS	ATTENDED COURSES	AVERAGE
86-0001	Bernd	Dresden	{(Calculus1, B, 86-1), (Alg, A, 87-9), (Sets, A, 87-5),...}	4,5
85-2738	Uwe	Pirna	{(Calculus1, D, 85-18), (Alg, C, 86-3), (Calculus2, C,87-2), (Geom, B, 86-22),...}	2,1
85-7389	Ulf	Freital	{(Calculus1, D, 85-8), (Alg, A, 86-23), (Calculus2, B, 86-2), (Geom, B, 86-2),...}	3,2
85-7129	Joe	Freiberg	{(Calculus1, C, 85-3), (Alg, A, 85-3), (Calculus2, A, 86-12), (Geom, B, 86-2),...}	3,8
85-1111	Joe	Ilmenau	{(Calculus1, D, 85-11), (Alg, D, 87-3), (Calculus2, D, 86-1), (Geom, C, 88-2),...}	1,3

The following relation scheme can be used for this table;

STUDENT = (U,$\underline{D}$,dom) with

U = {IDNUMB, NAME, ADDRESS, ATTENDED COURSES, AVERAGE} ,

$\underline{D}$ = { set-of-identity-numbers, set-of-names, set-of-towns, set-of- triples-with-course-name-mark-number, set-of-average-grades}.

The function dom is obvious.

There are several known restrictions:

- each student has its own identity number;
- in each course each student gets its own number.

These both restrictions can be used to distinguish all rows in the table. There are two minimal keys: {IDNUMB} and {ATTENDED COURSES}. Because of its structure, the attribute ATTENDED COURSES has a very high complexity. It can be used for the search of tuples but in most cases the utilization of the IDNUMB as search attribute would be more efficient. If in this university example other relation schemes are added to the presented relation scheme which are connected with the one presented then the modeling of the association between those schemes would be more complex and, therefore, it would be inefficient if the attribute ATTENDED COURSES were used instead of the attribute IDNUMB.

Given a set U of attributes, a subset X of U , a set S of subsets of U , the set of natural numbers including 0 , and a function
$g : U \longrightarrow N'$ (called complexity measure of U).
Then $g(X) = \Sigma_{A<-X}\, g(A)$ is called the **complexity** of X .
An element Y of S is called **g-shortest** if there does not exist an element Z of S with $g(Z) < g(Y)$.

By S(g) we denote the set of all g-shortest elements of S .
Relation schemes with constant (non-constant) functions g are called uniform (non-uniform) relation schemes.

It is easy to see that the g-shortest key can not be considered as a generalization of the notation of the minimal keys. Between the minimal keys there is selected a set of keys with the minimal complexity. Any system of g-shortest keys is a Sperner system. But there are Sperner systems which are not a set of g-minimal keys. In /LUOS 78/ and /BDFS 84/ it is proved that the following problem is NP-complete:
Given a relation scheme and an integer $m > 1$, decide whether there exists a key of cardinality less than m.
Consequently, if $NP \neq P$, then the time complexity of any algorithm that determines 1-minimal keys, is exponential.

By $S_r(g)$ we denote the set of all g-shortest elements of a key set S_r and by $s_r(g)$ its cardinality.

Corollary 4.4.10. Let RS = (U,D,dom) be a relation scheme, r a relation on RS, S the set of all keys of r , S_r the set of all minimal keys of r and g be a complexity measure of U . Then $S_r(g) = S(g)$, $S_r = S$, $S(g) = S_r$.
There exist relations on RS for which the inclusions are proper.

Lower and upper bounds for $s_r(g)$ are provided in /THAL 84/. The most interesting set of functions g is the set G^+ of functions g with $g(A_i) \neq g(A_j)$

for $i \neq j$. The other cases can be considered as a set of different cases: different constant function for different sets $X_1,\dots,X_m$ of attributes where the sets X_i are pairwise disjoint. Using this partition we consider the case that the clustered complexity function $g' : \{X_1,\dots,X_m\} \to N$ is now a function from G^+. We introduce the following functions:

$s(g) = \max_r \; s_r(g)$,

$s(G') = \max_{g \in G'} \; s(g)$ for sets G' of complexity measures from G of U .

Using the functions g_1, g_2, g_3 with

$g_1(A_i) = 2^i$,

$g_2(A_i) = 3^{i/2}$,

$g_3(A_i) = i$, for i , $1 \leq i \leq n$,

by the definitions and a recursion formula for g_3 /THAL 84/, we get

<u>Corollary 4.4.11.</u> 1. For complexity measures g of U , $|U| = n$, it holds $1 \leq s(g) \leq Fak(n)$.

2. $s(g_1) = 1$,

$s(g_2) = 2^{n/2}$,

$s(g_3) \geq 2^n / n^2$.

Our next aim is to prove

<u>Theorem 4.4.12.</u> $s(G^+) = \dfrac{2^n (1 - o(1))}{\sqrt{(\pi/6)\, n^3}}$.

We need some preparations for the proof. From number theory /KNOS 24/ we take that functions g with $s(g) = s(G^+)$ must be regular. W.l.o.g. we consider a subclass G^* of G^+ , the class of equidistant functions g with the property $g(A_i) - g(A_{i-1}) = c$ for some c and any i , $2 \leq i \leq n$.

Lemma 1. 1. Given two equidistant functions g, g' from G^+ . Then $s(g) = s(g')$.

2. Let g be a function from G^+ . There exists an equidistant function g' in G^* such that $s(g) \leq s(g')$.

Proof. 1. This assertion is immediate.

2. W.l.o.g. we consider only functions g from G^{+} with $g(A_i) < g(A_{i+1})$ for $1 \leq i \leq n-1$. We prove the assertion by induction. For $n = 2$ the assertion is obvious. Let n be a fixed number. Now we assume that for a fixed function g there is no equidistant function $g' \in G^{*}$ such that $s(g) \leq s(g')$. Let S_r be a key system with $s(g) = s_r(g)$.

Define $S_1 = \{K \in S_r \;/\; A_n \notin K\}$,

$S_2 = \{K-\{A_n\} \;/\; K \in S_r,\; A_n \in K\}$.

By the induction hypothesis for $g' = g|_{U'}$, $U' = U-\{A_n\}$ there is an equidistant function g'' such that $s(g') \leq s(g'')$. It follows that there is an equidistant function g^{+} in G^{*} such that $g^{+}|_{U'} = g''$ and $s(g) \leq s(g^{+})$.

That is a contradiction.

W.l.o.g. we can consider for $s(G^{+})$ the function g_3 of Corollary 4.1.11.

Define

$k(m,n) = \{ (n_1,\ldots,n_l) \mid l \leq l,\; 1 \leq n_1 < n_2 < \ldots < n_l \leq n,\;\; n_1+n_2+\ldots+n_l = m \}$

and $\underline{s}(m,n) = | k(m,n) |$.

Obviously, the following recursion formulas hold:

$\underline{s}(m,n) = \underline{s}(m-1,n-1) + \underline{s}(m,n-1)$,

$\underline{s}(1,n) = \underline{s}(\frac{1}{2}\; n\;(n+1)\;,\; n) = 1$

$\underline{s}(0,n) = \underline{s}(m,n) = 0 \quad$ for $m > n(n-1)/2$.

<u>Corollary 4.4.13.</u> $\underline{s}(n\;(n+1)\;/4\;,\; n) = s(g_3)$.

Now we define independent random variables rv_k with two-point distribution for $k = 1,2,\ldots,n$:

$$rv_k = \begin{cases} k \\ 0 \end{cases}$$

and consider the distribution of $Sr_n = \Sigma_{i=1}^{k}\; rv_i$.

<u>Corollary 4.4.14.</u> $P(Sr_n = (n(n+1))/4\;) \cong \frac{1}{2}n\; s(g_3)$ for the probability $P(Sr_n{=}m)$.

For the expectation ESr_n and the variance DSr_n of Sr_n we get

$$M_n = ESr_n = \sum_{k=1}^{n} Erv_k = \sum_{k=1}^{n} \frac{k}{2} = \frac{n(n+1)}{4} ,$$

$$B^2_n = DSr_n = \sum_{k=1}^{n} Drv_k = \frac{n(n+1)(2n+1)}{24} \sim \frac{1}{12} n^3 \ (n \to \infty) .$$

We shall say that the sequence $\{Sr_n\}$ satisfies a local limit theorem iff

$\sup_m |B_n P(Sr_n = m) - f(x_{nm})| \to 0 \ (n \to \infty)$

where $B_n x_{nm} = m - M_n$, $B_n z_n = Sr_n - M_n$, and f is the standard normal distribution density.

Put

$$al(a,q,N) = \frac{1}{q^2} \sum_{-q/2<r\leq q/2} r^2 P(a\, rv^{\sim}_k = r (\mathrm{mod}\ q) ,\ |rv_k| \leq N) \qquad (+)$$

where $rv^{\sim}_k = rv_k - rv'_k$ symmetrized random variable, rv'_k is a random variable independent of rv_k which has the same distribution as rv_k, relatively prime integers a,q with $a \leq \frac{q}{2}$ and $1 < q \leq 2N$.

Now we shall use the approach of /SETH 88/.

In /MITA 66/ the following is proved: If the distribution function of the sum of unboundedly increasing number of random variables converges to the standard normal distribution function,

$$\text{if} \quad z_n \xrightarrow{D} N(0,1) , \text{ and} \qquad (1)$$

$$N_n \exp\{ -\tfrac{1}{2} \min_{a,q} \sum_{k=1}^{n} al_k(a,q,N_k)\} \to 0 \quad (n \to \infty) , \qquad (2)$$

where N_n is selected such that $\underline{\lim}_{n\to\infty} \frac{1}{B^2_n} \sum_{k=1}^{n} \int_{|x|\leq N_n} x^2 \, dF_{rv\,k}(x) = 1 > 0$, (3)

then the sum satisfies a local limit theorem.

Let $N_n = n$. Then we get

$$1 = \lim_{n\to\infty} \frac{1}{B^2_n} \sum_{k=1}^{n} D\, rv^2_k = 1 > 0 ,$$

$$P(rv_k = k) = P(rv^{\sim}_k = -k) = \frac{1}{4} , \quad P(rv^{\sim}_k = 0) = \frac{1}{2} .$$

Summation of (+) over representatives of q yields $|rv^{\sim}_k| \leq n$ for $1\leq k \leq n$. Observe that if $rv^{\sim}_k = 0$ then $r = 0$ and this summand can be eliminated and that if $rv^{\sim}_k = k$ then $a\,k = r_k + q\, l_k$ for the unique representative of q.

Thus

$$al_k(a,q,N) = \frac{1}{q^2} \sum_{-q/2<r\leq q/2} r^2 P(a\, rv^{\sim}_k = r (\mathrm{mod}\ q)) =$$

$$= \frac{1}{4q^2} (r^2_k + r^2_{-k}) \geq \frac{1}{4q^2} r^2_k .$$

From number theory it is known that if {x} forms a full system of representatives of q then {ax} form a full system of representatives.

Now $\mathrm{lambda}_n \geq \min_{a,q} \sum_{k=1}^{n} al_k(a,q,N) \geq \min_q \frac{1}{4q^2} \sum_{k=1}^{n} r^2_k$.

Assume that $q = 2m$.(For odd q the proof is analogous)

Let $0 < \mathrm{alph} < \frac{1}{2}$. If $al_n \leq m \leq n$ then

$$\sum_{k=1}^{n} r^2_k \geq \sum_{k=1}^{n} k^2 \geq c\, m^3 \geq c\, \mathrm{alph}^3 n^3$$

for the full system of representatives $r_k = -(m-1),\ldots 0,1,\ldots,m$ and therefore

$$\mathrm{lambda}_n \geq \min_q \frac{1}{4q^2} c\, \mathrm{alph}^3 n^3$$

$$\geq (c\, \mathrm{alph}^3 n^3):(4\, \mathrm{alph}^2\, 4\, n^2) = b\, n \quad , \quad b > 0 .$$

If $1 < m < \mathrm{alph}\, n$ then the full system of representatives $\{r\}^m_{-(m-1)}$ is contained in {1,2,...,n} at least n/q times. Consequently we get

$$4 q^2 \ \mathrm{lambda} \geq \min_{a,q} \sum_{k=1}^{n} r^2_k \geq \min_{a,q} [n/q] \sum_{k=1}^{n} k^2$$

$$\geq \min_{a,q} (\frac{n}{q} - 1) \sum_{k=1}^{n} k^2 \geq \min_q (\frac{n}{q} - 1) \sum_{k=1}^{n} k^2 .$$

Now $\mathrm{lambda}_n \geq \min_q (n- 2\, \mathrm{alph}\, n)c = c(1- 2\, \mathrm{alph})n = b\, n > 0$ for $b > 0$.

We conclude that (2) holds because $\mathrm{delta}_n = n \exp\{ -\frac{1}{2} \mathrm{lambda}_n\}$

$$\leq n \exp\{ -\frac{b}{2} n\} \longrightarrow 0 \text{ for } n\to \infty .$$

Combining corollary 4.1.14 , lemma 1 and the properties of Sr_n we get

$$s(g_3) = 2^n P(Sr_n=n(n+1)/4) = (2^n : B_n) \ f(x_{n\,(n(n+1)/4)}) \ \sim_{n\to\infty}$$

$$\sim_{n\to\infty} 2^n : (\sqrt{2\pi\ n(n+1)(2n+1):24}) =$$

$$= 2^n : (\sqrt{(\pi:6)n^3(1 + (3:(2n)) - (1:(2n^2)))}) \ \sim_{n\to\infty}$$

$$\sim_{n\to\infty} 2^n : \sqrt{(\pi:6)\, n^3} .$$

The proof of theorem 4 is complete.

It is of interest to compare this result with $s(g_4) \sim 2^n :(\sqrt{\frac{\pi}{2}})\, n$ for $g_4(A) = 1$ for $A \in U$.

Using an integral local theorem and a central limit theorem /SETH 88/ we obtain the further result that for some constant c

$$| s(G^+) - 2^n : (\sqrt{(\pi:6)n^3(1 + (3:(2n)) + (1:(2n^2)))}) | \leq c : \sqrt{n} .$$

4.5. ARMSTRONG DATABASES

Armstrong relations are of practical use as they can effectively code the information on the dependencies they satisfy and they may be used as a design tool and a source of sample data for program testing. They are a partial solution to the problem of helping a designer to think about what dependencies should be included. This design aid then provides the database designer with an Armstrong relation, that is, a "sample relation" that obeys just those dependencies that are logical consequences of those that he has put in. The database designer needs not explicitly think about a specific dependency and whether it is a consequence of the dependencies he put in or not; rather, by inspecting the Armstrong relation, and thinking about what it says, he simply noticed that a dependency failed or succeeded. They help the designer and the database administrator select the dependencies to be included or to be considered. This verification by example has always been an alternative to formal deduction. Historically for example, the Babylonians wrote $(3 + 5)^2 = 3^2 + 2*3*5 + 5^2$, from which they immediately concluded all the other instances of the general formula $(x + y)^2 = x^2 + 2*x*y + y^2$. The use of "generic" examples can be observed occasionally by various degrees of explicitness. A concept closely related to Armstrong relations in traditional mathematics is the free algebra in equational logic or the generic algebras in universal algebra.
Unfortunately, there are limitations to this approach: That is a minimal-sized Armstrong relation for a set of keys can be of exponential size in the number of attributes.

Given a class K of dependencies from L(DRS) for DRS = $RS_1, RS_2, \ldots RS_m$ and a subset C of K . A database $\underline{r} = (r_1, \ldots r_m)$ is called <u>Armstrong database</u> for C in K if for all $d \in K$ $\underline{r} \models\models d$ if and only if $C \models d$.

A class K is called <u>Armstrong class</u> iff for any sound subset C of K there exists an Armstrong database for C in K .

For uni-relational classes K of dependencies, a relation r of an Armstrong database (r) is called <u>Armstrong relation</u>. If the class K is given by context, r is called Armstrong relation. For different special classes of dependencies there can be introduced special notations.

Given a Sperner set S of subsets of U, i.e. $X,Y \in S$ then $X \not\subseteq Y$ and $Y \not\subseteq X$. A relation r is called **Armstrong relation** for S if $S_r = S$.

Obviously, a class K is Armstrong iff from $\{\alpha_1,\ldots,\alpha_k\} \models \beta_1 \vee \beta_2 \vee \ldots \vee \beta_l$ follows that there is an β_i such that already $\{\alpha_1,\ldots,\alpha_k\} \models \beta_i$ for $\alpha_1,\ldots,\alpha_k, \beta_1,\ldots,\beta_l \in K$.

If an Armstrong database exists for any sound subset in a class K an utility criterion for Armstrong databases is the complexity of such structures for subsets of K.

The first example of application of theorem 4.4.7 to Armstrong relations concerns the number of elements of an Armstrong relation of key systems.
Now, let $a_K(S)$ denote the minimum number of tuples in Armstrong relations of S, where S is a Sperner set.
Let

$$a_K(n) = \max_{S\text{ -Sperner set on }U} a_K(S)$$

<u>Corollary 4.5.1.</u> $a_K(S) \geq \sqrt{2\,|S^{-1}|}$ where by S^{-1} is denoted the set of antikeys of S.

It should be noticed that the estimation $a_K(S) \geq \sqrt{2\,|S|}$ is not valid.
For instance, let $U = \{1,2,3,4,5,6\}$
$S = \{\{1,2\},\{1,3\},\{1,4\},\{1,5\},\{1,6\},\{2,3\},\{2,4\},\{2,5\},\{2,6\},\{3,4\},\{3,5\},\{3,6\},$
$\{4,5\},\{4,6\},\{5,6\}\}$. We get $|S| = 15$ and $\sqrt{2\,|S|} > 5$.
We construct the following relation r over U:

r	1	2	3	4	5	6
	1	1	1	1	1	1
	1	2	2	2	2	2
	2	1	3	2	3	3
	3	3	1	3	2	3

We see that $S_r = S$. Therefore $a_K(S) \leq 4$.

<u>Theorem 4.5.2.</u> /DEGY 81/ $\frac{1}{n^2} \binom{n}{\lfloor n/2 \rfloor} < a_K(n) \leq \binom{n}{\lfloor n/2 \rfloor} + 1$.

Proof. For the proof of theorem 4.4.2 it is clear that the number of elements of a mimimum-sized Armstrong relation is at most $\binom{n}{\lfloor n/2 \rfloor} + 1$. For the proof of the lower bound, we start by two trivial observations.
1. Let r be a relation over U with m tuples. Then there is a relation r' on RS such that r' uses not more than m symbols and $E(r) = E(r')$. Remember that $E(r)$ is the equality set of r.

2. Let r be a relation on RS with m tuples and $m' > m$. Then there is a relation r' over U with m' tuples such that $K(r) = K(r')$.
By 1. and 2. the number of Sperner systems which may be represented as sets of minimal keys of a relation with m tuples is no more than m^{nm}.

Hence $a_K(n)^{n*a_K(n)} > 2^{\binom{n}{[n/2]}}$ which implies $a_K(n) > \frac{1}{n^2}\binom{n}{[n/2]}$.

Let S^n_k denote the family of all k-element subsets of an n-element set U and let $a(n,k) = \max_{S \in S^n_k} a(S)$.
In /DEKA 83/, an estimation is given :

$c_1 \; n^{\frac{k-1}{2}} \leq a_1(n,k) \leq c_2 \; n^{\frac{k-1}{2}}$ where c_1, c_2 do not depend on n.

Using the inequality $\binom{p}{2} \geq \binom{m-n}{1}$ of proof of theorem 4.4.7, we get the following lower bound

<u>Corollary 4.5.3.</u> $a(n,k) > \sqrt[4]{\frac{(k-1)}{9n(n-k+1)}} \left(\frac{2n}{k-1}\right)^{(k-1)/2}$.

This estimate is of interest in the context of the following consequence of the definition of keys: If X is a key of a k-valued relation r then $|X| \geq \log_k |r|$ (e.g. $k^{|X|} \geq |r|$).

As already mentioned, there is an equivalence between monotone Boolean functions and sets of keys. Any monotone Boolean function f with n variables can be represented in the following way:

$$f = \bigwedge_{i=1}^{k} D_i = \bigvee_{j=1}^{t} K_j$$

where $D_i = x_{i1} \vee \ldots \vee x_{ik(i)}$, $K_j = x_{j1} \wedge \ldots \wedge x_{jl(j)}$ for $1 \leq i \leq k$, $1 \leq j \leq t$.
Let $\underline{S}(f) = \{\{A_{j1},\ldots,A_{jl(j)}\} = U \mid x_{j1} \wedge \ldots \wedge x_{jl(j)} \leq f\}$
and S_r be the set of all keys of a relation r,
where for Boolean functions $\leq$ denotes the logical smaller or equal relation.
Obviously, the function $\bigvee_{(A_{j1},\ldots,A_{jl(j)}) \in S_r} (x_{j1} \wedge \ldots \wedge x_{jl(j)})$ is a monotone Boolean function for any relation r.
Applying theorem 4.4.2 we obtain

__Corollary 4.5.4.__ Let be $f = \bigwedge_{i=1}^{k} D_i$ a monotone n-ary Boolean function. Then there is a k-valued relation r with $S_r = \underline{S}(f)$ and $| r | \leq k+1$.

Note that there are monotone functions f such that no 2-valued relation r exists with $S_r = \underline{S}(f)$. The function $f = x_1x_2 \vee x_3$ is an example. If $\{A_1,A_2\}$ is a minimal key for $r = \{(a,b,0), (c,d,1)\}$ then $a = c$ or $b = d$ and consequently, $\{A_1\}$ or $\{A_2\}$ is a minimal key. But $\{A_1,A_2\}$ and $\{A_3\}$ are minimal keys of the 3-valued relation $r = \{(0,0,0), (0,1,1), (1,0,2)\}$.

__Theorem 4.5.5.__ Let $f = D_1 \wedge \ldots \wedge D_m$ be a monotone function, let $D_i = x_{i1} \vee \ldots \vee x_{ik(i)}$ be disjunctions for any i , $1 \leq i \leq m$, and let $k = 1 + \max\{k(1), k(2),\ldots,k(m)\}$. If a code $C = \{1,\ldots,q\}^n$ of distance k and with m elements exists then there is a (2q)-valued relation r with $|r| = 2t$ and $\underline{S}(f) = S_r$.

Proof. Let $f = D_1 \wedge \ldots \wedge D_m$. Suppose, a q-valued code $C = \{c_{11}\ldots c_{1n}, \ldots , c_{m1}\ldots c_{mn}\}$ is of the Hamming-distance k .
We construct the tuples t_i of r as follows:

$t_i(A_j) = c_{ij}$ for any $1 \leq i \leq m$, $1 \leq j \leq n$,

$$t_{i+m}(A_j) = \begin{cases} c_{ij} + q & \text{if } x_j \leq D_i \\ c_{ij} & \text{otherwise} \end{cases} \qquad 1 \leq i \leq m, \quad 1 \leq j \leq n .$$

Now we get for the Hamming-distance dis of elements of r :
$dis(t_i,t_{i+m}) < dis(t_i,t_j) \leq dis(t_i,t_{j+m})$ and
$dis(t_i,t_{i+m}) < dis(t_{i+m},t_{j+m})$ for any $i \neq j$, $1 \leq i,j \leq m$.

Consequently, $\bigvee_{t_i(A_m)=t_{i+m}(A_m)} x_m \leq \bigvee_{t_{i'}(A_m)=t_{i''}(A_m)} x_m$
for $(i',i'') \in \{ (i,j), (i,j+m), (i+m,j+m) \}$ $1 \leq i,j \leq m$, $i \neq j$.
We obtain now

$$\bigwedge_{i=1}^{m} \Big(\bigvee_{t_i(A_m)=t_{i+m}(A_m)} x_m \Big) = K_1 \vee \ldots \vee K_o = D_1 \wedge \ldots \wedge D_m .$$

Using this proof, a 2-valued relation r on $U = \{1,\ldots,2n\}$ with $S_r =$

$\{ X\{n+1,\ldots,2n\} \mid X \in \underline{S}(f) \}$ can be easily constructed for arbitrary monotone Boolean functions f .

Now we shall consider classes of generalized functional dependencies being Armstrong sets. Here we use the approach of /BEBL 85/ and /THAL 84/.

For a set $C = \{(f_i,g_i) \mid 1 \leq i \leq m\}$ of generalized functional dependencies and a relation r on RS ,

$$E^*(r) = \{ \sigma' \mid \sigma(t,t') \leq \sigma' , \quad t,t' \in r \}$$

$$T(C) = \{ \sigma \mid (f \dashrightarrow g)(\sigma) = 1 \text{ for all } (f,g) \in C \} .$$

Lemma 4.5.6. r is Armstrong for C iff $E^*(r) = T(C)$.

Proof. Given r and C . We know that

$r \mid\mid\!\!== C$ iff $\{t,t'\} \mid\mid\!\!== C$ for all $t,t' \in r$

iff $(f\dashrightarrow g)(\sigma(r,r')) = 1$ for all $t,t' \in r$ and all $(f,g) \in C$

iff $E(r) \subseteq T(C)$

iff $E^*(r) \subseteq T(C)$.

Now let $\sigma \in T(C) - E(r)$. For (f_C,g_C) constructed by theorem 4.1.4 (the root of C) we get now the contradiction

$r \mid\mid\!\!=\!\!\not\; (f_C,g_C)$ and $C \mid\!= (f_C,g_C)$. Therefore, $T(C) \subseteq E(r)$ for Armstrong relations r for C .

A generalized functional dependency (f,g) is called positive if $f(0,\ldots,0) \leq g(0,\ldots,0)$. Let be $GFDEP^+$ the set of positive generalized functional dependencies.

Theorem 4.5.7. The sets $GFDEP^+$, FDEP, SFDEP, KFDEP, DFDEP $\cap$ $GFDEP^+$, WFDEP $\cap$ $GFDEP^+$, MFDEP $\cap$ $GFDEP^+$ of positive generalized functional dependencies, functional dependencies, strong functional dependencies, key dependencies, positive dual functional dependencies, positive weak functional dependencies and positive monotone functional dependencies resp. are Armstrong sets. The sets GFDEP, DFDEP, WFDEP, MFDEP of generalized functional dependencies, dual functional dependencies, weak functional dependencies and monotone functional dependencies resp. are not Armstrong sets.

Proof. 1. Let $C \subseteq GFDEP^+$ and $T(C) = \{(0,\ldots,0), \sigma_1,\ldots,\sigma_m \}$.
For each $\sigma_j = (\sigma_{j1},\ldots,\sigma_{jn})$ define a relation $r_j = \{t_j,t_j'\}$ by

$t_j(i) = 2j$

$$t_j'(i) = \begin{cases} 2j & \text{if } \sigma_{ji} = 1 \\ 2j-1 & \text{if } \sigma_{ji} = 0 \end{cases} \qquad 1 \leq i \leq n \ , \ 1 \leq j \leq m \quad .$$

Note $\sigma(t_j, t_j') = \sigma_j$ for $1 \leq j \leq m$ and for $i \neq j$

$\sigma(t_j, t_i) = \sigma(t_j', t_i) = \sigma(t_j, t_i') = \sigma(t_j', t_i') = (0, \ldots, 0)$.

So $E^*(r) = T(C)$ and consequently, r is an Armstrong relation for C .

2. Let $C \subseteq$ FDEP. We show that $T(C)$ contains a least element. In order to show that it is sufficient to show that if $\sigma, \sigma' \in T(C)$ then $\sigma \hat{} \sigma' = (\sigma_1 \hat{} \sigma_1', \ldots, \sigma_n \hat{} \sigma_n') \in T(C)$ as well.

If $\sigma \hat{} \sigma' \notin T(C)$ then for some $X \text{-->} Y \in C$ $(K_X \text{-->} K_Y)(\sigma \hat{} \sigma') = 0$ where by K_Z is denoted the conjunction of x_i for $A_i \in Z$.

From this follows $\sigma \notin T(C)$ or $\sigma' \notin T(C)$, i.e. a contradiction.

Now using the construction in 1. we get an Armstrong relation r for C with $\sigma(t_j, t_i) = \sigma(t_j', t_i) = \sigma(t_j, t_i') = \sigma(t_j', t_i') = \sigma_m$ for the least element of $T(C)$.

3. Because of subsets of Armstrong sets are Armstrong sets the other sets are Armstrong sets.

4. /BEBL 85/ Now consider that

$C = \{ \emptyset \text{-->}^D \{A_1, A_2\} \} \subseteq$ DFDEP (and $C \subseteq$ WFDEP).

Suppose $E^*(r) = T(C) = \{ \sigma \mid \sigma \geq \sigma_1 \text{ or } \sigma \geq \sigma_2 \}$ with $\sigma_1 = (1,0,\ldots,0)$, $\sigma_2 = (0,1,0,\ldots,0)$ and $t_1, t_2, t_3, t_4 \in r$ such that $\sigma(t_1, t_2) = \sigma_1$ and $\sigma(t_3, t_4) = \sigma_2$. Now, as σ_1 , σ_2 are the only minimal elements of $T(C)$ either $\sigma(t_1, t_4) \geq \sigma(t_1, t_2)$ or $\sigma(t_1, t_4) \geq \sigma(t_3, t_4)$ and so without loss of generality assume $\sigma(t_1, t_4) \geq \sigma(t_1, t_2)$. So,

$t_1(A_1) = t_2(A_1) = t_4(A_1) \neq t_3(A_1)$.

On the other hand, either $\sigma(t_2, t_3) \geq \sigma(t_1, t_2)$ or $\sigma(t_2, t_3) \geq \sigma(t_3, t_4)$.

If $\sigma(t_2, t_3) \geq \sigma(t_1, t_2)$ then $t_3(A_1) = t_1(A_1)$, a contradiction.

If $\sigma(t_2, t_3) \geq \sigma(t_3, t_4)$ then $t_1(A_2) \neq t_2(A_2) = t_3(A_2) = t_4(A_2)$.

But now it follows that none of $\sigma(t_1, t_3) \geq \sigma(t_1, t_2)$ and $\sigma(t_1, t_3) \geq \sigma(t_3, t_4)$ holds, a contradiction.

5. Since if a subset of a set is not an Armstrong set the set is not an Armstrong set, the other sets are not Armstrong sets.

There are also other techniques to construct Armstrong databases. Another important method is presented in the proof of lemma 4.2.8. In /FAG 82/, four different techniques to construct Armstrong relations and the limitations of these techniques are discussed: disjoint union (technique was first suggested for FD's and MVD's in /BEFH 77/); agreement sets (lemma 4.2.8); direct products of relations (/ARM 74/, /FAG 80/); chase method (see chapter 3.2.3). In /THAL 84/, another technique, called direct union, is used for constructing Armstrong databases for JD's.

An easy extension of lemma 4.2.8 (see for example /DEGY 83/, /BDFS 84/) leads to

Theorem 4.5.8. /DEGY 83/, /BDFS 84/ There is a constant c such that for each set C of FD's involving n attributes, there is an Armstrong relation for C with less than

$\binom{n}{\lfloor n/2 \rfloor}(1 + c/\sqrt{n})$ tuples. For each positive integer n, there is a set C of FD's involving n attributes such that each Armstrong relation for C contains more $\frac{1}{n^2}\binom{n}{\lfloor n/2 \rfloor}$ tuples.

The first proof of the lower bound was given in /DEME'80/. This results follows directly from theorem 4.5.2 because of $a_K(n) \leq a\ (n)$.

Using lemma 4.2.8 we get (/THAL 84/) also

Corollary 4.5.9. 1) $\frac{1}{n^2}\binom{n}{\lfloor n/2 \rfloor} \leq a\ (n) \leq \binom{n}{\lfloor n/2 \rfloor} + 1$.

2) $a_{GFDEP^+}(n) \leq 2^{n-1} - 1$.

In /BDFS 84/ it is also shown that the complexity of finding an Armstrong relation, given a set of FD's, is precisely exponential in the number of attributes. In order to prove that, the authors point out a set C of functional dependencies so that the number of tuples in a minimal Armstrong relation is exponential, not only in the number of attributes, but also in the number of functional dependencies. In /DETH 87/ a stronger result is given. The time complexity of finding an Armstrong relation for a given Sperner system S is exactly exponential in the number of elements of S and in the number of elements of U. The algorithm used in /DETH 87/ has a good average case behavior.

Now we will find out how large the domain size in an Armstrong relation must be, that is, we consider the valueness of Armstrong relations.

Theorem 4.5.10. There is a constant c such that every minimal Armstrong relation of $C \subseteq$ FDEP with n attributes contains less than

$$\binom{n}{\lfloor n/2 \rfloor} (1 + c / \sqrt{n})$$ distinct values in each column. There is a set of FD's such that each Armstrong relation for this set is k-valued for some $k > \frac{1}{2n^2}\binom{n}{\lfloor n/2 \rfloor}$.

Proof. The upper bound follows from theorem 4.5.8., since the number of distinct values in each column is bounded by the number of tuples.
We consider now the lower bound. Let $m = n-1$ and $k = \lfloor m/2 \rfloor$.
By theorem 4.5.8, where m plays the part of n, we know that there is a set C of FD's (over the first m attributes $A_1,\ldots,A_{n-1}$) such that each Armstrong relation for C' contains more than $m^{-2} \binom{m}{k}$ tuples. Let C contain C', along with exactly one more FD $\{A_n\} \rightarrow \{A_1,\ldots,A_{n-1}\}$.
Thus the new FD reveals that the new attribute A_n forms a key. Each Armstrong relation r for C contains more than $m^{-2} \binom{m}{k}$ tuples, since the projection of r onto the first m attributes is an Armstrong relation for C' with as many tuples as r. Since $\{A_n\}$ is a key, every tuple has a distinct A_n-value.
Thus by $\frac{1}{(n-1)^2}\binom{n-1}{\lfloor \frac{n-1}{2} \rfloor} \geq \frac{1}{n^2}\binom{n}{\lfloor n/2 \rfloor}$, the A_n-column contains more than $\frac{1}{n^2}\binom{n}{\lfloor n/2 \rfloor}$ values.

We note that by a simple modification /BDFS 84/ of the proof of theorem 4.5.8, it can be proved that for each constant k, we have $a_{FDEP}(n) > \frac{1}{(n-k)^2}\binom{n}{\lfloor n/2 \rfloor}$ for n sufficiently large. Using this bound the lower bound of theorem 4.5.10 can be improved.

4.6. DEGENERATED MULTIVALUED DEPENDENCIES

Let us consider the class of degenerated multivalued dependencies which is a subclass of generalized functional dependencies. This class was introduced in /ARDE 80/ and also considered in /SDPF 81/ . Given the relation scheme $RS = (U , \underline{D}, dom)$ where $U = \{A_1,\ldots,A_n\}$, subsets X, Y, Z of U. The propositional dependency $X \rightarrow Y \vee Z$ is called <u>degenerated multivalued dependency</u>. If $XYZ = U$ the degenerated multivalued dependency is called <u>full</u>.
Any degenerated multivalued dependency can be represented by a generalized func-

tional dependency. We associate with each attribute A_i in U a Boolean variable x_i and denote by K_V the conjunction of all Boolean variables associated with the attributes of the set V . Then the Boolean function corresponding to a degenerated multivalued dependency X -> Y v Z is $K_X \rightarrow (K_Y \vee K_Z)$.

__Corollary 4.6.1.__ The degenerated dependency X--> Y v Z is valid in a relation r on RS iff for all tuples t , t' from r with t(X) = t'(X) is valid t(Y) = t'(Y) or t(Z) = t'(Z) . The functional dependency is a special degenerated multivalued dependency. Key dependencies are special full degenerated multivalued dependencies.

With corollary 4.6.1 functional dependencies X --> Y can be considered as degenerated multivalued dependencies X --> Y v Ø . Another equivalent degenerated multivalued dependency is X --> Y v Y .

Also for degenerated multivalued dependencies theorem 4.1.4 can be applied for the characterization of the implication problem. Therefore, we obtain directly an algorithm for the solution of the implication problem.

Let us now consider some derivation rules.

(DMD0) For any X,Y, Z $\subseteq$ U XY --> Y v Z

For subsets X, Y, Z, V, W , X', Y', Z', V', W' of U :

(DMD1)	$\dfrac{X \rightarrow Y \vee Z}{X \rightarrow Z \vee Y}$	(commutability)
(DMD2)	$\dfrac{X \rightarrow Y}{X \rightarrow Y \vee Z}$	(first augmentation)
(DMD2')	$\dfrac{X \rightarrow Y \vee Z}{XWV \rightarrow YV \vee Z}$	(second augmentation)
(DMD3)	$\dfrac{X \rightarrow Y \vee Z}{X \rightarrow (Y-X) \vee Z}$	(branch minimalization)
(DMD4)	$\dfrac{X \rightarrow Y \ ,\ Y \rightarrow V \vee W}{X \rightarrow V \vee W}$	(first transitivity)
(DMD4')	$\dfrac{X \rightarrow Y \vee Z \ ,\ Y \rightarrow V}{X \rightarrow V \vee Z}$	(second transitivity)
(DMD5)	$\dfrac{X \rightarrow Y \vee Z \ ,\ X \rightarrow V \vee W}{X \rightarrow (Y \cap V) \vee (ZW)}$	(decomposition)
(DMD6)	$\dfrac{X \rightarrow YY' \vee Z \ ,\ XZ \rightarrow Y \vee Y'}{X \rightarrow Y \vee Y'Z}$	(branch interchange)

(DMD7) $\dfrac{X \dashrightarrow Y \vee Z}{X \dashrightarrow Y \cap Z}$ (branching)

(DMD7') $\dfrac{X \dashrightarrow Y \vee Z}{X \dashrightarrow (Y-Z) \vee (Z-Y)}$ (branch subset)

(DMD7") $\dfrac{X \dashrightarrow V \ , \ X \dashrightarrow Y \vee Z}{X \dashrightarrow YV \vee Z}$ (branch union)

(DMD8) $\dfrac{XY \dashrightarrow Z \ , \ X \dashrightarrow V \vee W}{X \dashrightarrow V}$ if $V \subseteq ZY$ and $V \cap Y \subseteq W$ (first mixing with FD's)

(DMD8') $\dfrac{XX' \dashrightarrow YY' \vee ZZ' \ , \ X \dashrightarrow Y}{XX' \dashrightarrow Y'}$ (second mixing with FD's)

(DMD8") $\dfrac{X \dashrightarrow VV' \vee WW' \ , \ XVV' \dashrightarrow VW}{X \dashrightarrow W}$ (third mixing with FD's)

Using theorem 4.1.4 we obtain

__Corollary 4.6.2.__ The rules (DMD0),...,(DMD8") are sound.

It is easy to see that the rules (DMD5), (DMD8') and (DMD8") can be derived using the other rules.
Let us define for the set of FD's and full degenerated multivalued dependencies the rules (FDMD i) by appending to the rule (DMD i) the condition that full degenerated multivalued dependencies and FD' are allowed only.
Let Γ_{FDMD} be the formal system containing (FDMD0), (FD0), (FDMD1), (FDMD2), (FDMD2'),(FDMD3), (FDMD 4), (FDMD4'), (FDMD6), (FDMD7), (FDMD7'),(FDMD8). Using the same proof as considered in chapter 4.2 we obtain

__Corollary 4.6.3.__ The system Γ_{FDMD} is sound and complete for the implication of full degenerated multivalued dependencies and functional dependencies.

For the incompleteness of (DMD0) - (DMD8") let us consider

__Example 4.6.4__. Let be given for a relation scheme RS = (U , $\underline{D}$, dom) where
$U = \{A_0,\dots,A_{k-1}\}$ the set C and d be defined as follows:
$\{\{A_1\} \dashrightarrow \{A_2\} \vee \{A_0\} \ , \ \{A_2\} \dashrightarrow \{A_3\} \vee \{A_0\} \ , \dots , \ \{A_{k-2}\} \dashrightarrow \{A_{k-1}\} \vee \{A_0\} \ ,$
$\{A_{k-1}\} \dashrightarrow \{A_1\} \vee \{A_0\} \ \}$ and
d = $\{A_1\} \dashrightarrow \{A_{k-1}\} \vee \{A_0\}$.
Using theorem 4.1.4 we get $C \models d$. All rules considered above are 1-ary or 2-

ary. By theorem 3.1.2, lemma 3.4.2 and lemma 3.4.3 there is no k-ary axiomatization of the class of degenerated multivalued dependencies.

5. JOIN DEPENDENCIES

The decomposition of a relation in a relational database management system is a central issue that has been extensively studied during the last decennium. There are many reasons for decomposing a relation. The most important seem to be

- smaller relations are easier to understand, to quest and to compute;
- no orthogonal, redundant information should be included in an unique relation;
- in distributed databases different components can be located in different sites.

The decomposition may have some disadvantages. Decomposition by normalization possibly makes it easier to update the database, but it clearly makes a database more difficult to query if the join is needed for the evaluation of the answer since the join operation can be considerably expansive with respect to computations to be performed.

First, multivalued dependencies (/FAG 77/,/ZANI 76/) were studied. They are used for decomposing a relation in two components. In /RISS 78/, join dependencies (JD) are introduced as a generalization of multivalued dependencies. Hierarchical dependencies were introduced by /DELO 73/. Several special cases of JD's were studied in detail, hitherto.

Given a relation scheme $RS = (U , \underline{D} , dom)$ where $U = \{ A_1,\ldots,A_n\}$.
For pairwise disjoint subsets X, Y, Z, $Y_1,\ldots$ of U the following join dependencies are

(XY , XZ) <u>binary join dependency</u> (or <u>multivalued dependency</u> $X \to\to Y$),
$(Y_1,\ldots,Y_m)$ <u>full cross</u>,
$(XY_1,\ldots, XY_m)$ <u>generalized multivalued dependency</u> (or <u>full hierarchical dependency</u> , denoted by $X : Y_1|Y_2|\ldots|Y_m$),
$(XY_1, XY_2,\ldots, XY_{2m}, Y_1Y_2, Y_3Y_4,\ldots Y_{2m-1}Y_{2m})$ <u>mixed dependency</u>,
$(XY_1, XY_2,\ldots, XY_m, Y_1Y_2, Y_2Y_3,\ldots Y_{m-1}Y_m)$ <u>codependency</u>,
$(XY_1, XY_2,\ldots, XY_{m1}, Y_1Y_{11},\ldots, Y_1Y_{11},\ldots,Y_{m1}Y_{m1\ 1},\ldots,Y_{11}Y_{111},\ldots$
$\ldots,Y_{m1\ m2\ldots m(s-1)}Y_{m1m2\ldots ms})$ <u>s-tree dependency</u>,
(XY, XZ, YZ) <u>mutual dependency</u>, <u>contextual join dependency</u>,
$(Y_{12}Y_{13}\ldots Y_{1m},Y_{12}Y_{23}\ldots Y_{2m},\ldots, Y_{1m}Y_{2m}\ldots Y_{(m-1)m})$ <u>graphical dependency</u>.

For not necessary disjoint subsets $Y_{12}, Y_{13},\ldots,Y_{1m},\ldots,Y_{(m-1)m}$ of U the JD $(Y_{12}Y_{13}\ldots Y_{1m},Y_{12}Y_{23}\ldots Y_{2m},\ldots, Y_{1m}Y_{2m}\ldots Y_{(m-1)m})$ is called <u>generalized mutual dependency</u>.
For sets V, W the union of these sets is denoted by VW .

The last six sorts of dependencies can be better understood by their hypergraphical representation. For any join dependency $d = (X_1,\ldots X_n)$ on U there can be defined the hypergraph $H(d) = (U, \{X_1,\ldots,X_n\})$. Mixed dependencies are represented by hypergraphs with a root X which represent a tree structure where neighboring odd and even leaves are connected. In hypergraphs of codependencies the neighboring leaves are connected. The hypergraph of s-tree dependency has a tree structure of height s . The hypergraph of graphical dependencies is represented by a graph structure. In hypergraphs of generalized mutual dependencies there are no nodes $A \in U$ which are only in one component X_i .

Given the following relation scheme RS = (U , <u>D</u>, dom) where U = {A,B,C,D,E,F,G,H,I,J,K}. Then the following examples of different dependencies can be considered:

({A,B,C,D,E,F},{A,B,C,D,G,H,I,J,K} binary join dependency;

({A,B,C},{D,E,F},{G,H},{I,J},{K}) full cross;

({A,B,C},{A,B,D},{A,B,E,F,G,H},{A,B,I,J,K}) generalized multivalued dependency;

({A,B,C,D},{A,B,C,E},{A,B,C,F},{A,B,C,G},{A,B,C,H},{A,B,C,I},{A,B,C,J},{A,B,C,K}, {D,E},{F,G},{H,I},{J,K}) mixed dependency;

({A,B,C,D},{A,B,C,E},{A,B,C,F},{A,B,C,G},{A,B,C,H},{A,B,C,I},{A,B,C,J},{A,B,C,K}, {D,E},{E,F},{F,G},{G,H},{H,I},{I,J},{J,K}) codependency;

({A,B},{A,H},{B,C},{B,E},{H,G},{H,I},{C,D},{E,F},{G,J},{I,K}) 3-tree dependency;

({A,B,C,D},{A,B,E,F},{A,B,G,H},{A,B,I,J,K},{C,D,E,F},{C,D,I,J,K},{E,F,G,H}, {H,I,J,K}) graphical dependency;

({A,B,C,D,E,F},{A,B,G,H,K},{D,G,H,I},{C,E,G,J},{I,J,K}) generalized mutual dependency.

The class of FD's, MVD's, mutual dependencies, full hierarchical dependencies, mixed dependencies and codependencies is also class of <u>root dependencies</u>.

Now we consider only the transformation with projection and join. There are also other transformations with projection where the reconstruction map is not necessarily the join /FAVA 84/. Today, it is not known whether there is an effective test for the necessity of join or other operations. It depends on the set of integrity constraints C . If from C follows a jd $d = (X_1,\ldots,X_n)$ then the reconstruction map for the transformation with projection via d is the join. The inverse holds, too. Let us consider the following example.

Given a relation scheme RS = ({1,2,3},<u>D</u>,dom), X = {1,2}, Y = {1,3}, dom(A) = {0,1} for $A \in \{1,2,3\}$. Let us consider the relation r = {(0,0,0),(1,0,1)}. Then

$r \not\models (X,Y)$ but $r = r(X) + r(Y)$.

In /AABM 82/, the following connection is proven. Given three relation schemes $RS = (U, \underline{D}, dom)$ where $U = \{A_1,\ldots,A_n\}$, $RS' = (U', \underline{D}, dom')$ where $U' = \{B_1,\ldots,B_m\}$, $RS'' = (U'', \underline{D}, dom'')$ where $U'' = \{C_1,\ldots,C_l\}$, $X = U' \cap U''$, $Y = U' - X$, $Z = U'' - X$, $XYZ = U$, and a set C of functional dependencies on U and a set of dependencies C' on $U'U''$, respectively. Let C'' the set of functional dependencies which is implied by C' .

The two schemes (RS,C) and (RS',RS'',C') are equivalent if and only if $U = U'U''$ $C \models C'$, $C'' \models C$, and

$\forall x \forall y \exists z (P'(x,y) \dashrightarrow P''(x,z))$, $\forall x \forall z \exists y (P''(x,z) \dashrightarrow P'(x,y)) \in C'$

and $C \models X \dashrightarrow Y$ or $C \models X \dashrightarrow Z$.

For the remaining part of this chapter, we assume that a fixed natural number n and a fixed relation scheme $RS = (U, \underline{D}, dom)$ where $U = \{A_1,\ldots,A_n\}$ are given.

In chapter 5.1., we consider the properties of the most important subclass of join dependencies. The class of binary join dependencies (multivalued dependencies) is axiomatizized. In chapter 5.2, full hierarchical dependencies are explored. Some properties of acyclic join dependencies are presented in chapter 5.2. In chapter 5.3. we present some results on the class of join dependencies.

5.1. MULTIVALUED DEPENDENCIES AND BINARY JOIN DEPENDENCIES

In this chapter we show the usefulness of methods presented in chapter 4.2. Our first aim is the axiomatization of the class $JDEP_2$ of binary join dependencies.

We say that a jd (X,Y) <u>is stronger than</u> a jd (V,W) (denoted by $(X,Y) \leq (V,W)$) if $X \subseteq V$ and $Y \subseteq W$ or $Y \subseteq V$ and $X \subseteq W$.

Let us consider the following formal system Γ_{JD2} .

Axiom (A1) $(\emptyset,U)$

Rules

(21) $\dfrac{(X,Y)}{(V,W)}$ if $(X,Y) \leq (V,W)$

(22) $\dfrac{(X,Y)\ ,\ (V,W)}{(X \cap V,\ W)}$ if $X \cap Y \subseteq V$, $Y \subseteq W$.

This system generalizes the formal system Γ_{JD2}. of /ARDE 80/ where instead of rule (22) the rule

(23) $\dfrac{(X,Y)\ ,\ (V,W)}{(X \cap V\ ,\ W)}$ if $V \cap W = Y$

is used.

In analogy to chapter 4.2., we introduce some characterizations.

<u>D_2-characterization</u> . Let $C \subseteq JDEP_2$. Then we say that C satisfies the D_2-characterization if for any $(X,Y) \in JDEP_2 - C$ there is an $E \subseteq U$ such that

(1) $X \cap Y \subseteq E$, $X \not\subseteq E$ and $Y \not\subseteq E$;

(2) if $(X',Y') \in C$ and $X' \cap Y' \subseteq E$ then $X' \subseteq E$ or $Y' \subseteq E$.

<u>D'_2-characterization</u>. Let $C \subseteq JDEP_2$. Then we say that C satisfies the D'_2-characterization if there is a natural number k and an indexed set of subsets of U $\underline{E} = \{E_{ij} \mid 1 \le i < j \le k\}$ such that

(1) $(X,Y) \in C$ iff for any $E_{ij} \subseteq \underline{E}$ if $X \cap Y \subseteq E_{ij}$ then $X \subseteq E_{ij}$ or $Y \subseteq E_{ij}$;

(2) for any i,j,l , $1 \le i < j < l \le k$, any different pairs $(m,p), (q,s) \in \{(i,j), (i,l), (j,l)\}$ $E_{ij} \cap E_{il} \cap E_{jl} = E_{mp} \cap E_{qs}$.

Remember that for a class K the set C , $C \subseteq K$, is <u>K-closed</u> if for any $\alpha \in K$ $C \models \alpha$ implies $\alpha \in C$. For a class K and a formal system Γ_K a set C , $C \subseteq K$, is (K, Γ_K)-full if for $\alpha \in K$ $C \vdash_{\Gamma K} \alpha$ implies $\alpha \in C$.

<u>**Theorem 5.1.1.**</u> Let $C \subseteq JDEP_2$. then the following are equivalent:

1) C is $(JDEP_2, \Gamma_{JD2})$-full.
2) C satisfies the D_2-characterization.
3) C satisfies the D'_2-characterization.
4) C is $JDEP_2$ -closed.

Lemma 1 - lemma 6 prove theorem 5.1.1.

<u>Lemma 1.</u> If C is $JDEP_2$-closed then C is $(JDEP_2, \Gamma_{JD2})$-full.

Proof. (A1) and (21) are very easy to prove and are left to the reader to be proved.

For (22) without loss of generality there is a partition $\{Z_1,Z_2,Z_3,Z_4,Z_5,Z_6\}$ of U such that $(X,Y) = (X_1,X_2) = (Z_1\ Z_2\ Z_3\ Z_4\ ,\ Z_1\ Z_5\ Z_6)$ and

$(V,W) = (Y_1,Y_2) = (Z_1\ Z_2\ Z_3\ Z_5\ ,\ Z_1\ Z_2\ Z_4\ Z_5\ Z_6)$.

Let r be a relation on RS and t, t' be tuples in r with $t(Z_1\ Z_2) = t'(Z_1\ Z_2)$. We want to find a tuple $t"$ in r with $t"(Z_1\ Z_2\ Z_3) = t(Z_1\ Z_2\ Z_3)$ and $t"(Z_1\ Z_2\ Z_4\ Z_5\ Z_6) = t'(Z_1\ Z_2\ Z_4\ Z_5\ Z_6)$ assuming $r \models C$. Since $C \vdash_{\Gamma_{JD2}} (X_1,X_2)$, we can find t^* in r with

$t^*(Z_1\ Z_2\ Z_3\ Z_4) = t(Z_1\ Z_2\ Z_3\ Z_4)$ and $t^*(Z_1\ Z_5\ Z_6) = t'(Z_1\ Z_5\ Z_6)$.

Because $t^*(Z_1\ Z_2\ Z_5) = t'(Z_1\ Z_2\ Z_5)$ there is a tuple t^+ in r with $t^+(Y_1) = t(Y_1)$ and $t^+(Y_2) = t'(Y_2)$. The relation r satisfies the requirements, since $t^+(Y_1 \cap X_1) = t^*(Y_1 \cap X_1) = t(Y_1 \cap X_1)$.

<u>Lemma 2</u>. If C is $(JDEP_2\ ,\ \Gamma_{JD2})$-full, then C satisfies the D_2-characterization.

Proof. Let C be a $(JDEP_2\ ,\ \Gamma_{JD2})$-full family of binary join dependencies. Suppose that $(X\ V\ ,\ V\ Y) \notin C$ for some partition $\{X,Y,V\}$ of U . By finiteness of U there exists a maximal subset E of U such that $V \subseteq E$ and E maximal for $(X\ V,\ V\ Y)$, that is $(X\ E,Y\ E) \notin C$ and for E' , $E' \neq E$, $E \subseteq E'$, $(X\ E',\ Y\ E') \in C$. We should show that this E meets the conditions in the D_2-characterization. First of all, if we had $X \subseteq E$ then we would have $(E,E\ Y) \notin C$ and hence $(U,\emptyset) \notin C$, in contrary to the assumption. Hence $X \not\subseteq E$, and, similarly , $Y \not\subseteq E$.

Suppose next that $(V'\ X',V'\ Y') \in C$, $V' \subseteq E$ for some partition $\{X',Y',V'\}$ of U . Now suppose that $X' \not\subseteq E$ and $Y' \not\subseteq E$. From $X" = X' - E$, $Y" = Y' - E$,

$(E\ X",E\ Y") \in C$ we get $(E\ X\ X",\ E\ X"\ Y),\ (E\ X\ Y",E\ Y\ Y") \in C$.

From $(E\ X",E\ Y"),\ (E\ X\ Y",E\ Y\ Y") \in C$ we get

$(E\ (X \cap X"),E\ Y\ Y"),\ (E\ Y\ Y",E\ X) \in C$.

From $(E\ Y",E\ X"),\ (E\ X\ X",E\ Y\ X") \in C$ we get $(E\ (X \cap Y"),E\ Y\ X") \in C$.

From $(E\ Y\ X",E\ (X \cap Y")),\ (E\ Y\ Y",E\ X) \notin C$ we get $(E\ Y,\ E\ X) \in C$, in contrary to the assumption. Then C satisfies the D_2-characterization.

Let $\underline{X} = \{X_1,\dots,X_m\}$ be a set system. Then $\underline{X}$ is a Φ-system, if for any i, j, k, l, $1 \leq i,j,k,l \leq m$, $i \neq j$, $k \neq l$ $X_i \cap X_j = X_k \cap X_l$.

Lemma 3. If C satisfies the D_2-characterization then C satisfies the D'_2-characterization.

Proof. For any $(X,Y) \notin C$ take an $E(X,Y) \subseteq U$ guaranteed by the D_2-characterization. List these $E(X,Y)$'s as $E_2,\dots,E_k$. For $1<j\leq k$ let $E_{1j} = E_j$ and for $1<i<j\leq k$ let $E_{ij} = E_i \cap E_j$.
The requirement (1) of the D'_2 characterization holds by
$\{E_2,\dots,E_k\} \subseteq \{E_{ij} \mid 1\leq i<j\leq k\}$.
To prove (2) of the D'_2-characterization let $1\leq i<j<l\leq k$. There are two cases:
1. $i = 1$. Then $E_{ij} = E_j$, $E_{il} = E_l$, $E_{jl} = E_j \cap E_l$. Thus
$\{E_{ij},E_{il},E_{jl}\}$ is a Φ-system.
2. $i > 1$. Then $E_{ij} = E_i \cap E_j$, $E_{il} = E_i \cap E_l$, $E_{jl} = E_j \cap E_l$. Thus
$\{E_{ij},E_{il},E_{jl}\}$ is a Φ-system.

For elements t, t' from r, $M = (\underline{D},r)$ let
$E(t,t') = \{ A \in U \mid t(A) = t'(A) \}$ and
$E(r) = \{E(t,t') \mid t,t' \in r, t \neq t' \}$.

Lemma 4. Let r be a relation on RS and let t, t', t'' be different elements of r. Then $\{E(t,t'),E(t,t''),E(t',t'')\}$ forms a Φ-system.

We left to the reader to examine that lemma 4 holds.

Lemma 5. Let $\underline{E} = \{E_{ij} \mid 1\leq i<j\leq k\}$ such that for each i,j,l, $1\leq i<j<l\leq k$, $\{E_{ij},E_{il},E_{jl}\}$ is a Φ-system. Then there is a relation r on Rs with $E(r)=\underline{E}$.

Proof. We construct by induction the tuples $t_1,\dots,t_k$ of r for $\underline{D} = \{ \mathbb{N}' \}$, $U = \{A_1,\dots,A_n\}$, $dom(A) = \mathbb{N}'$ where by $\mathbb{N}'$ is denoted the set of natural numbers including 0.
Let $t_1(A) = 0$ for $A \in U$, and assume that $m < k$ and the tuples $t_1,\dots,t_m$ have been defined such that for each $1\leq i<j\leq m$ $E(t_i,t_j) = E_{ij}$ holds.
We construct t_{m+1} as follows:

$$t_{m+1}(A) = \begin{cases} t_i(A) & \text{if } A \in E_{i(m+1)} \text{ for some } i,\ 1 \le i \le m ; \\ \max\{t_i(A) \mid 1 \le i \le m\} + 1 & \text{else} . \end{cases}$$

Then it is clear that for $1 \le i \le m$, $E(t_i, t_{m+1}) = E_{i(m+1)}$ and hence the induction step works. Let $r = \{t_1, \ldots, t_k\}$. Then obviously $E(r) = \underline{E}$ holds.

<u>Lemma 6.</u> Let $C \subseteq JDEP_2$ satisfy the D'_2-characterization. Then there is a database $M = (\underline{D}, r)$ on RS with $C = \{d \in JDEP_2 \mid r \mid\mid== d\}$.
Conversely, if r is relation on RS then $\{d \in JDEP_2 \mid r \mid\mid== d\}$ satisfies the D'_2-characterization.

Proof. Let $\underline{E} = \{E_{ij} \mid 1 \le i < j \le k\}$ show that C satisfies the D'_2-characterization. Then the requirement (2) of the D'_2-characterization and lemma 5 imply that there is such a relation r with $E(r) = \underline{E}$. By the D'_2-characterization it is obvious that $C = \{d \in JDEP_2 \mid r \mid\mid== d\}$.
Conversely, if r is a relation on RS, then by $r = \{t_1, \ldots, t_k\}$, $E_{ij} = E(t_i, t_j)$, $\underline{E} = \{E_{ij} \mid 1 \le i < j \le k\}$ the set $\{d \in JDEP_2 \mid r \mid\mid== d\}$ satisfies the D'_2-characterization.

There are also known other formal systems /THAL 84/.

Formal system $\Gamma_{JD2''}$.
Axiom (A1)
Rules (21)

$$(24) \quad \frac{(X_1, X_2) \ , \ (Y_1, Y_2)}{(Y_1 \cap (X_1 Y_2), Y_2 X_2)}$$

Formal system $\Gamma_{JD2'''}$.
Axiom (A1)
Rules (21)

$$(25) \quad \frac{(X_1, X_2), (Y_1, Y_2), (Z_1, Z_2)}{(V_1, V_2)}$$

where $V_1 = (X_1 \cap (X_2 Y_1 Z_1))(Y_1 \cap (Y_2 Z_2))$, $V_2 = (X_2 \cap (X_1 Y_2 Z_1))(Y_2 \cap (Y_1 Z_2))$.

Formal system Γ_{JD2IV} /BFH 77/ .

Axiom (A1)

Rules (21)

$$(26)\quad \frac{(X_1,X_2),(Y_1,Y_2)}{(Y_1 \cap (X_1Y_2),Y_2X_2)} \quad \text{if } X_1 \cap X_2 \subseteq Y_2 \text{ and } Y_2 \subseteq (X_1 \cap X_2)Y_1 \ .$$

Formal system Γ_{JD2V}.

Axiom (A1)

Rules (21)

$$(27)\quad \frac{(X_1,X_2),(Y_1,Y_2)}{(X_1 \cap Y_1, X_2Y_2)} \quad \text{if } X_1 \cap X_2 = Y_1 \cap Y_2 \ .$$

It is easy to prove that the formal systems Γ_{JD2} , $\Gamma_{JD2'}$, $\Gamma_{JD2''}$, $\Gamma_{JD2'''}$, and Γ_{JD2IV} are equivalent. From $C \vdash_{\Gamma JD2} d$ follows $C \vdash_{\Gamma JD2IV} d$.

For $d = (X_1,X_2)$, $d' = (Y_1,Y_2)$, $d'' = (Z_1,Z_2) \in {}_2$ let $d''' = (X'_1,X'_2)$ be some dependency with $Z_1 \cap Z_2 \subseteq X'_2$, $Z_2 \subseteq X'_1$ and $d''' \geq d$.
Then the following tree using the rule (24) leads to the result of the rule (25).

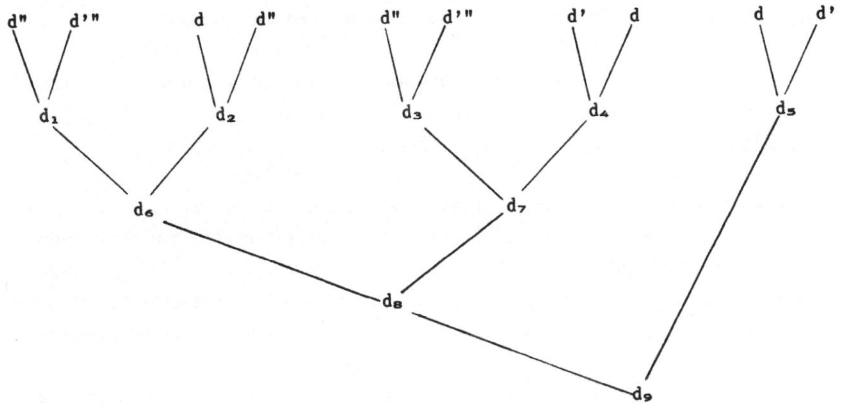

For automated computation, the formal system $\Gamma_{JD2'"}$ and $\Gamma_{JD2"}$ are the most convincing ones. The rules (24) and (25) are both rules without conditions.
Now we can summarize the previous results in theorem 5.1.2. which shows the equivalence of the introduces formal systems.

Theorem 5.1.2. Let C be a system of binary join dependencies and d be a binary join dependency. Then the following are equivalent:

1) $C \models d$.
2) $C \vdash_{\Gamma_{JD2}} d$.
3) $C \vdash_{\Gamma_{JD2'}} d$.
4) $C \vdash_{\Gamma_{JD2"}} d$.
5) $C \vdash_{\Gamma_{JD2'"}} d$.
6) $C \vdash_{\Gamma_{JDIV}} d$.

In contrary to the assumptions in the literature the formal system Γ_{JD2v} is not complete (Corollary 5.1.3.).

Corollary 5.1.3. The formal system Γ_{JD2v} is sound but not complete.

Proof. Since the system $\Gamma_{JD2''}$ is sound and the rule (27) is a special case of the rule (24) , the system Γ_{JD2v} is sound. A rule of the form

$$\frac{(X,X'),(Y,Y')}{(Z,Z')} \quad \text{condition1} \qquad \text{or} \qquad \frac{(Y,Y')}{(Z,Z')} \quad \text{condition2}$$

is called root cardinality reducing if there exist sets X,X', Y, Y' or Y,Y' which fulfill the conditions such that $|Z \cap Z'| < \max(|X \cap X'|,|Y \cap Y'|)$ resp. $|Z \cap Z'| < |Y \cap Y'|$.

The rules (22), (23), (24), (25), (26) are root cardinality reducing but (27) is not root cardinality reducing. Therefore, the system Γ_{JD2v} cannot be complete.

Using the equivalence between multivalued dependencies and binary join dependencies we get the following formal system for multivalued dependencies /BFH 77/, /BISK 78/.

Formal system Γ_{MVD} .

Axioms (A2) XY ->-> Y X,Y ⊆ U

Rules (11) $\frac{X \to\to Y}{X \to\to Z}$ if XYZ = U and $Y \cap Z \subseteq X$

(12) $\frac{X \to\to Y}{XWZ \to\to YZ}$

(13) $\frac{X \to\to Y\ ,\ Y \to\to Z}{X \to\to Z-Y}$

Formal system $\Gamma_{MVD'}$.

Axioms (A2)

Rules (11)

(12)

(14) $\frac{X \to\to Y\ ,\ X' \to\to Y'}{X(X'-Y) \to\to Y'-Y}$

Formal system $\Gamma_{MVD''}$.

Axioms (A3) $\emptyset$ ->-> U

Rules (12)

(13)

Corollary 5.1.4. The systems Γ_{MVD} , $\Gamma_{MVD'}$, $\Gamma_{MVD''}$ are sound and complete for the implication of MVD's.

There are also known different other rules which can be used for faster derivation:

(12') $$\frac{X \rightarrow\rightarrow Y,\ X \rightarrow\rightarrow Z}{X \rightarrow\rightarrow YZ}$$

(12") $$\frac{X \rightarrow\rightarrow Y,\ X \rightarrow\rightarrow Z}{X \rightarrow\rightarrow Y \cap Z}$$

(12'") $$\frac{X \rightarrow\rightarrow Y,\ X \rightarrow\rightarrow Z}{X \rightarrow\rightarrow Y - Z}$$.

Analogously to corollary 5.1.3, it can be proven that these rules cannot replace the rule (13) or the rules (12) in complete formal systems.

A special problem is the problem of transitively specified MVD's. Two transitively specified MVD's are shown often to impose a semantically unnatural constraint for relations. In /KATY 79/ the following property of transitively specified dependencies is shown to be valid:
If X,Y,Z are non-empty disjoint sets of attributes and X->-> Y, Y->->Z hold in r , then $r[x,Z] = r[x',Z]$ for all X-values x,x' such that $r[x,Y] \cap r[x',Y] \neq \emptyset$ and $r[y,Z] = r[y',Z]$ for all Y-values y , y' such that $r[y,X] \cap r[y',X] \neq \emptyset$.
The constraints X ->-> Y , Y ->-> Z are semantically unnatural constraints because neither X-values nor Y-values can determine a set of Z-values independently. If additionally X -> Y or Y -> X holds in r then the semantical problem of transitively specified MVD's does not occur. If neither X -> Y nor Y -> X holds, then any decomposition of r[XYZ] causes a serious problem under update operations. The implied MVD X ->-> Z cannot be maintained independently in r[XY] and r[YZ] (similar in r[XY], r[XZ] the MVD Y ->-> Z) under update operations.

Without proof we present the following sound and complete formal system $\Gamma_{FD,JD2}$ for the implication of functional and binary join dependencies. The proof of theorem 5.1.1. can be used for the proof of soundness and completeness.

Formal system $\Gamma_{FD,JD2}$

Axioms (A1)

(A4) $X \rightarrow X$ for $X \subseteq U$

Rules (21)

(22)

(23)

(15) $\dfrac{X \rightarrow Y\ ,\ Y \rightarrow Z}{XVW \rightarrow ZW}$

(16) $\dfrac{X \rightarrow Y}{(XY,\ X(U-Y))}$

(17) $\dfrac{(X,Y)\ ,\ X \rightarrow Z}{X \cap Y \rightarrow Y \cap Z}$.

Using the proof of theorem 5.1.1. we get

Corollary 5.1.5. For any C , $C \subseteq JDEP_2$, there exists an Armstrong relation r with $|r| \leq 2^n$.

Now we want to give a combinatorial characterization of those sets which are of minimal cardinality with respect to the property that they imply all dependencies of a given $JDEP_2$-closed set.

Let $N^*(C)$ denote the minimal size of a minimal generating subset C' of C , i.e. $C' \models C$ and $C' - \{d\} \not\models d$ for each $d \in C'$.

Let $N^*_2(n)$ denote the maximum size of $N^*(C)$ for $JDEP_2$-closed sets C in a database with n attributes.

Theorem 5.1.6. $n^{-1/2}\ 2^{n-1} \leq N^*_2(n) \leq 2^n\ (1 - 1/(n+1))$.

Proof. The upper bound follows from corollary 4.2.12. For the proof of the lower bound we use a property of the presented formal systems. A formal system Γ of binary join dependencies is called root cardinality preserving if for any rule

$\dfrac{(X_1,Y_1),\ldots,(X_m,Y_m)}{}$ of Γ the following property is valid

$(V,W) \qquad |V \cap W| \geq \min (|X_1 \cap Y_1|,\ldots,|X_m \cap Y_m|)$.

Obviously, the system Γ_{JD2} is root cardinality preserving.

Now let $C = \{(X,Y) \in {}_2 \mid |X \cap Y| = [n/2] \}$.

For any set C' with $C' \models C$ we get

$$|C'| = \binom{n}{\frac{n}{2}}$$

because of Γ_{JD2} is root cardinality preserving.

Binary join dependencies or multivalued dependencies and functional dependencies can be represented by special Boolean functions. This representation is based on the similarity of semantical behavior of multivalued dependencies and degenerated multivalued dependencies. We associate with each attribute A_i in U a Boolean variable x_i and denote by K_X the conjunction of all Boolean variables associated with the attributes of the set X (see also chapter 4). Then the Boolean function corresponding to a FD or a binary JD or a MVD is defined as follows:

$X \rightarrow Y$ corresponds to $K_X \rightarrow K_Y$,

$X \rightarrow\rightarrow Y$ corresponds to $K_X \rightarrow (K_Y \vee K_{U-Y})$ and

(X,Y) corresponds to $K_{X \cap Y} \rightarrow (K_X \vee K_Y)$ where $K_\emptyset = 1$.

Theorem 5.1.7./SDPF 81/ Let F_C be the set of Boolean functions (resp. f_α the Boolean function) corresponding to the set of functional, multivalued and binary join dependencies (resp., a FD, MVD of binary JD α). Then from C follows α iff $\bigwedge_{f \in F_C} f \leq f_\alpha$.

The proof of this theorem is omitted and can be easily reconstructed by theorem 4.1.4. and theorem 4.1.6. In /SDPF 81/ it is stated in contrary to theorem 5.2.6. that theorem 5.1.7. cannot be extended to known generalizations of MVD's.

For database logical design, normalization and effective algorithms, it is useful to utilize the full information on given relations. In a great number of applications, there is a requirement to allow violation of some MVD's, i.e. MVD's that are intended but do not hold in the relation.

The constraint

$\exists x \exists y \exists y' \exists z \exists z' \, (P(x,y,z) \wedge P(x,y',z') \wedge (\neg P(x,y,z') \vee \neg P(x,y',z)))$

is called <u>excluded multivalued constraint</u> and for

$X = \{A_i \in U \mid x_i \neq x\}$, $Y = \{A_i \in U \mid y_i \neq y\}$ and $Z = \{A_i \in U \mid z_i \neq z\}$ denoted by X ->/-> Y .

The axiomatization of MVD's and excluded multivalued constraints is found in /THAL 89/. The following formal system is sound and complete.

Formal system $\Gamma_{MVD,EMVC}$.

Axioms (A2) .

Rules (11)

(12)

(13)

(11) $\dfrac{X \text{ ->/-> } Y}{X \text{ ->/-> } Z}$ for $XYZ = U$, $Y \cap Z \subseteq X$

(121) $\dfrac{XWZ \text{ ->/-> } YZ}{X \text{ ->/-> } Y}$ if $Y \neq \emptyset$

(131) $\dfrac{X \text{ ->-> } Y \text{ , } X \text{ ->/-> } Z\text{-}Y}{Y \text{ ->/-> } Z}$

(132) $\dfrac{Y \text{ ->-> } Z \text{ , } X \text{ ->/-> } Z\text{-}Y}{X \text{ ->/-> } Y}$ if $Y \neq \emptyset$.

There are also other extensions of binary join dependencies, as for instance weak multivalued dependencies /JAES 82/. A formula

$\forall x \forall y \forall y' \forall z \forall z'$ (P(x,y',z') ∧ P(x,y',z) ∧ P(x,y,z') --> P(x,y,z))

is called <u>weak multivalued dependency</u>. The satisfaction of a certain set of weak multivalued dependencies yields a reasonable horizontal and vertical decomposition of a relation, even when the corresponding MVD is not satisfied. In /FIGU 85/, a complete and sound system for the implication of weak multivalued dependencies is presented.

5.2. FULL HIERARCHICAL DEPENDENCIES AND ACYCLIC JOIN DEPENDENCIES

Hierarchical dependencies are introduced by Delobel /DELO 73/. In /STPA 84/ based on the results of C. Delobel and M. Leonard the similarity between hierarchical dependencies and hierarchical data structures is illustrated. But for hierarchical dependencies a complete and sound formal system cannot exist /DEAD 85/ as it ensues from theorem 3.4.4. Therefore the class of full hierarchical dependencies is important as a generalization of multivalued dependencies, a class of a complete and sound formal system and because of its structure is a class of dependencies which is used in practice by estimates of /DEAD 85/ nearly by 25 % of practical applications. By HDEP, the class of full hierarchical dependencies is denoted.

For relations meeting certain full hierarchical dependency, it is useful to know equivalent conditions for control of satisfaction. The following theorem is a generalization of a theorem of V.P. Vashenko (1967)/VASH 78/.

Remember, that for a relation r on RS , a tuple t from r , $X \subseteq U$, the subset of tuples which are equal to t on X is denoted by r:t[X], formally

$r:t[X] = \{t' \in r \mid t'(X) = t(X)\}$.

Theorem 5.2.1. Given a relation scheme RS = (U , $\underline{D}$, dom) , a relation r on RS and the full hierarchical dependency $d = (XY_1, XY_2, \ldots, XY_m)$. The following are equivalent:

(1) $r \models d$.

(2) For any $t \in r$ $r:t[X] = \times_{i=1}^{m} (r:t[X])[Y_i] \times \{t(X)\}$.

(3) For any i , $1 \leq i < m$, $(r:t[X])[Y_i Y_{i+1}] = (r:t[X])[Y_i] \times (r:t[X])[Y_{i+1}]$ for any $t \in r$, $t_i \in (r:t[X])[Y_i]$, $1 \leq i \leq m$, $t[X], t_1, \ldots, t_m$ form a tuple of r.

Proof. 1. The equivalence of (1) and (2) follows by definition of JD's. Since $r = \bigcup_{t \in r} r:t[X]$ and $r:t[X] \models Y_i \rightarrow X$ if $r \models d$ we get

$r:t[X] = \ast_{i=1}^{m} (r:t[X])[XY_i] = \times_{i=1}^{m} (r:t[X])[Y_i] \times \{t(X)\}$. Conversely, if

$r:t[X] = \times_{i=1}^{m} (r:t[X])[Y_i] \times \{t(X)\}$ then $r:t[X] \models Y_i \rightarrow X$.

Since $r:t[X] \cap r:t'[X] = \emptyset$ for $r, r' \in r$ with $t(X) \neq t'(X)$ we get $r \models d$.

2. It is obvious that (3) follows from (2). It is sufficient to show that (1) follows from (3) . We must show that

$\ast_{i=1}^{m} r[X Y_i] \subseteq r$. Now let $t, t_1, \ldots, t_m$ such tuples as in condition (3) and forming a tuple in $\ast_{i=1}^{m} r[XY_i]$.

Because of $\{t_i\} \times \{t_{i+1}\} \subseteq (r:t[X])[Y_i Y_{i+1}]$ and $t[X], t_1, \ldots, t_m$ form a tuple of r we get $\{t[X]\} \times \{t_1\} \times \ldots \times \{t_m\} \subseteq r$.

Corollary 5.2.2. Any full hierarchical dependency $(XY_1,\ldots XY_m)$ is equivalent to a set C of binary join dependencies with $|C| =]\log_2 m[$ where the smallest natural number n with $n \geq k$ is denoted by $]k[$.

The proof is obvious when the soundness of the following rules is proved:

(HJD2) $\dfrac{d_1}{d_2}$ $d_1 \in HDEP,\ d_2 \in JDEP_2\ ,\ d_1 \leq d_2$ (for $d_1 = (X_1,\ldots,X_m)$ and $d_2 = (Y_1,Y_2)$, for any X_i it holds $X_i \subseteq Y_1$ or $X_i \subseteq Y_2$)

(H3) $$\frac{(XY_1\ ,\ldots\ ,\ XY_m)\ \ ,\ \ (XZ_1\ ,\ldots,\ XZ_k)}{(X(Y_1 \cap Z_1),\ldots,X(Y_1 \cap Z_k),X(Y_2 \cap Z_1),\ldots,X(Y \cap Z_k))}$$

The soundness of the first rule is obvious by monotony of join expressions. The soundness of the second rule follows directly from theorem 5.2.1.(2).

Denoting by $\langle k\rangle_1,\ldots\langle k\rangle_o$ the 1-ary dual representation of the number k we define the set C as follows:

$$\{\ (X \cup \bigcup_{\langle j\rangle_i=0} Y_j\ ,\ X \cup \bigcup_{\langle j\rangle_i=1} Y_j\ )\ \mid\ 0 \leq i < \log_2 m\ \}\ .$$

Now letter by letter with lemma 1 - lemma 6 from chapter 5.1., we can prove the following equivalence in somewhat puzzling analogy. First, we introduce a formal system for full hierarchical dependencies.

Formal system Γ_H .

Axiom (U)

Rules

(H1) $\dfrac{(X_1,\ldots,X_m)}{(Z_1,\ldots,Z_k)}$ if for some $(X_1,\ldots,X_m),(Z_1,\ldots,Z_k) \in$, for any i there is an j such that $X_i \subseteq Z_j$

(H2) $$\frac{(XY_1\ ,\ldots,\ XY_m)\ ,\ (VZ_1\ ,\ldots,\ VZ_k)}{(VZ_1,\ldots,VZ_{i-1},V(Z_i \cap Y_1),\ldots,V(Z_i \cap Y_m),VZ_{i+1},\ldots,VZ_k)}$$

if $Z_i \subseteq U - X$;

(H3)

This system is a subsystem of /BEVA 81/ (see also /BISK 78/) and can be obtained directly from its system using the property that only full hierarchical dependencies are required in derivation of hierarchical dependencies.

For set systems $\underline{F}$, $\underline{G}$ we write $\underline{F} \leq \underline{G}$ iff for every $G \in \underline{G}$ there is a

$F \in \underline{F}$ such that $F \subseteq G$.

Theorem 5.2.3. For any $C \subseteq$ HDEP, the following statements are equivalent:

(1) C is (HDEP , Γ_H)-full.

(2) C is HDEP-closed.

(3) There is a set $\underline{E}$ of subsets of U such that $(X_1,\ldots,X_m) \in C$ iff for all $E \in \underline{E}$ the property $X_1 \cap X_2 \subseteq E$ implies $\{X_1,\ldots,X_m\} \underline{\subset} \{E\}$.

In /THAL 84/ a direct proof for the equivalence of conditions (2) and (3) is presented which uses the following properties:

1) If C is JDEP-closed then $C \cap JDEP_2$ is $JDEP_2$-closed.

2) If C satisfies the condition (3) then $C \cap JDEP_2$ satisfies the D_2-characterization and therefore is $JDEP_2$-closed.

Now we notice that full hierarchical dependencies precisely behave like a certain fragment of propositional logic or a set of Boolean functions.

For the proof we use a semiorder relation $\geq$ in a subset HGFDEP of GFDEP (chapter 4.1.).

For any $d = (XY_1,\ldots,XY_m) \in$ HDEP let (f_d,g_d) the corresponding functional dependency with $f_d = K_X$ and $g_d = K_{U-Y_1} \vee\ldots\vee K_{U-Y_m}$.

Let $= \{(f_d,g_d) \mid d \in \text{HDEP}\}$. By corollary 4.1.11, we get that for any element of max(C) for a closed set $C \subseteq$ HGFDEP, there exists exactly one presentation $(f_1,g_1) \sqcup\ldots\sqcup (f_k,g_k)$ with u-irreducible elements of max(C). A functional dependency (f,g) is an element of a closed set C from GFDEP iff there exists an element (f',g') in max(C) such that $f' \geq f$ and $g' \leq g$ holds.
Therefore, by theorem 5.1.7. and corollary 5.2.2. we get three consequences.

Corollary 5.2.4. Let be $C \subseteq$ HDEP and $X \subseteq U$. Then, exactly one minimal full hierarchical dependency $d_{C,X} = (XX_1,\ldots,XX_k)$ exists such that:

1. $C \models d_{C,X}$.

2. A full hierarchical dependency $(Y_1,\ldots,Y_m)$ with $Y_1 \cap Y_2 = X$ is implied by C iff there $Y_i = X \cup \bigcup_{X_j \cap Y_i \neq \emptyset} X_j$ holds for any i , $1 \leq i \leq m$.

Corollary 5.2.4. can be proved directly using Γ_H .

Corollary 5.2.5. If there is no $(Y_1,\dots,Y_m) \in C$, $C \subseteq$ HDEP, with $Y_1 \cap Y_2 \subseteq X$ then holds $C \not\models (X_1,\dots,X_m)$ for $(X_1,\dots,X_m) \in$ HDEP if $X_1 \cap X_2 = X$.

Theorem 5.2.6. Let be $C \subseteq$ HDEP , $d \in$ HDEP . Then the following are equivalent:

1) $C \models d$.
2) $C \vdash_{\ulcorner H} d$.
3) $\{(f_{d'}, g_{d'}) \mid d' \in C\} \models (f_d, g_d)$.
4) $\bigwedge_{d' \in C} f_{d'} \to g_{d'} \leq f_d \to g_d$.

Some of the properties of full hierarchical dependencies can be generalized to other join dependencies. It can be denoted that JD's can be also represented by Boolean functions

$f_d(x_1,\dots,x_m)$ with $m < k^2 n/2$ for $d \in JDEP_k$ /THAL 84/ .

In literature, it is often claimed that in almost any "real world" situation, a single join dependency suffices, together with some functional dependencies, to define the legal databases that might be the uni-relational database some times. This assumption results in a great simplification in the algorithms required to interpret queries and to perform updates on the uni-relational database in a way that can be reflected in the actual relations of the database in a sensible manner.

But if the join dependency is a special one (later on called acyclic), then there is no ambiguity regarding interpretations of queries that connect two or more attributes. That is, there is a unique minimal set of relations that must be joined to get a relation by a set of attributes that includes the attributes involved in the query.

A join dependency is called <u>acyclic</u> iff it is equivalent to a set of binary join dependencies (or multivalued dependencies).

In /BFMY 83/ monotonous join expressions are considered. Given a relation scheme RS = (U , $\underline{D}$, dom) , a relation r on RS and an algebraic expression e . The algebraic expression e is monotonous with respect to r if for every subexpression (e_1 * e_2) of e the relations $e_1(r)$ and $e_2(r)$ are equal over the common attributes. Intuitively, e is monotonous with respect to r if no tuples are lost in taking any of the binary joins obtained by executing e(r) as dictated by the parenthesis.

Given a database scheme DS = (RS, C ∪{d}) where RS = (U , D , dom) and d is an acyclic join dependency. Then any DS-database (r) has a monotonous algebraic expression. Therefore, such databases provide a "space-efficient" manner for taking a join, so that no more tuples are evaluated in intermediate joins than in the final join.

There is an efficient algorithm for the test of acyclicity of a join dependency:

Graham's algorithm /BFMY 83/.

1. Given some JD $(X_1,..,X_m) \in$ JDEP .

2. For any i, $1 \le i \le m$, $X_i = \bigcup_{j=1,\ j \ne i}^{m} X_i \cap X_j$.

3. For any i , $1 \le i \le m$,

$$X_i = \begin{cases} \emptyset & \text{if there is an } X_j \text{ , } j \ne i, \text{ with } X_i \not\subseteq X_j \text{ or if there is} \\ & \text{an } X_j \text{ , } j>i, \text{ with } X_i = X_j \text{ ;} \\ X_i & \text{otherwise.} \end{cases}$$

4. Repeat 2. and 3. if there is some new result.

Theorem 5.2.7. A join dependency d is acyclic iff Graham's algorithm terminates for d with only empty sets.

In /GOTA 84/ the following connection is proven.

Theorem 5.2.8. The set C of binary join dependencies is equivalent to a join dependency d iff it is equivalent to a set C' of binary join dependencies with the following property: for every pair (X,Y) , (V,W) from C'

$X \subseteq V$, $W \subseteq Y$ or

$X \subseteq W$, $V \subseteq Y$ or

$Y \subseteq V$, $W \subseteq X$ or

$Y \subseteq W$, $V \subseteq X$.

In /BFMY 83/ there is characterization for sets of MVD's which are implied by a single join dependency. A necessary and sufficient property is the intersection property. A set C of MVD's has the intersection property if whenever $C \models X \to\to Z$ and $C \models Y \to\to Z$ with $X \cap Z = Y \cap Z = \emptyset$ then also $X \cap Y \to\to Z$ is implied by C .

R. Fagin had also introduced more restrictive types of acyclicity using special hypergraph properties. Various notions of acyclicity turn out to be useful for the design of universal relation interface. Adding in Grahams algorithm after 3. the step 3' then this algorithm will be a test of -acyclicity:

3'. $X_i = \emptyset$ if $|X_i| = 1$.

If $\{X_i \mid A_j \in X_i\} = \{X_i \mid A_k \in X_i\}$ for $k > j$ then delete in all X_i the attribute A_k .

5.3. THE CLASS OF JOIN DEPENDENCIES

The class of join dependencies is one of the most important classes for the database design theory. Therefore, its implication problem is of the greatest importance for the theory. Often, it is stated that dependencies given by a user are only of the classes of MVD's and FD's. There is a fundamental difference between constraints of the conceptual scheme - which are called in /THAL'88/ reality and design dependencies - and those constraints of the database scheme which are consequences of the way this scheme is obtained from the conceptual schemes - in /THAL'88/ called database constraints. But join dependencies are used by the database designer to decompose the database without loss of information. The proof procedure (see 3.2.3.) has an exponential worst-case running time. Moreover, in /MSY 81/ it is proved, that if C is a set of one join dependency and several functional dependencies, then testing whether C implies another join dependency d is NP-complete.

Theorem 3.1.3. cannot be used to find an axiomatization. It states only that a finite axiomatization exists for the class of join dependencies. In /THAL 84/ it is proved, that there is a set C of independent join dependencies (i.e. for a given $C \subseteq$ JDEP for any $d, d' \in C$, $d \neq d'$ $C-\{d\} \not\models d$) with more than

$c\ n^{1/4}\ 2^{2^{n-2}/\sqrt{n}}$ elements. Since the set of join dependencies consists of

more than $2^{2^{n}/\sqrt{n}}$ nonequivalent elements the axiomatization of JDEP with

theorem 3.1.3 is computational unfeasible.

Now we present two formal systems for join dependencies. We introduce some special notions for it :

Let $d = (X_1,\ldots,X_m)$, $d' = (Y_1,\ldots,Y_k)$ be set systems with subsets from U . We write $d \leq d'$ if for any i, $1 \leq i \leq m$, there is some j, $1 \leq j \leq k$, such that $X_i \subseteq Y_j$.

Let

$MANY(X_i,d) = X_i \cap (X_1 \ldots X_{i-1} X_{i+1} \ldots X_k)$, $1 \leq i \leq m$,

$ONCE(X_i,d) = X_i - MANY(X_i,d)$, $1 \leq i \leq m$,

$MANY(d) = MANY(X_1,d) \cup \ldots \cup MANY(X_m,d)$,

$ONCE(d) = U - MANY(d)$.

Using this notation, we get directly another characterization of the different introduced dependency classes. For instance, a join dependency d is a generalized mutual dependency if and only if $ONCE(d) = \emptyset$.

Formal system Γ_{JD} .

Axiom (A0) (U)

Rules (1) $\dfrac{d}{d'}$ if $d \leq d'$;

(2) $\dfrac{(X_1,\ldots,X_k) \ , \ (Y_1,\ldots,Y_m)}{(Z_1,\ldots,Z_k,Y_2,Y_3,\ldots,Y_m)}$

with $Z_i = MANY(X_i,(X_1,\ldots,X_k)) \cup (X_i \cap Y_1)$

for i , $1 \leq i \leq k$.

Formal system $\Gamma_{JD'}$ /BEVA 81/, /SCIO 82/ .

Axioms (A0)

Rules (1)

(2*) $\dfrac{(X_1,\ldots,X_k) \ , \ (Y_1,\ldots,Y_m)}{(X_1 \cap Y_1,\ldots,X_k \cap Y_1,Y_2,\ldots,Y_m)}$ if $MANY((X_1,\ldots,X_k)) \subseteq Y_1$

Corollary 5.3.1. Let $C \subseteq JDEP$, $d \in JDEP$. Then

$$C \vdash_{\Gamma JD} d \quad \text{iff} \quad C \vdash_{\Gamma JD'} d .$$

The formal system Γ_{JD} is very powerful.. Almost all known Hilbert-Type inference rules can be derived from Γ_{JD} .

Theorem 5.3.2. The system Γ_{JD} is JDEP-full.

Proof. For the proof we use the system Γ_{TD1} from chapter 3.3. This system is

JDEP-full. Therefore, we must show that the rules presented are equivalent to the rules of Γ_{TD1} and that for any derivation in Γ_{TD1} there is also a derivation in Γ_{JD} and vice versa.

1. Assume, that $\{d_1,d_2\} \vdash_{\Gamma JD} d$. Then a derivation $d'_1,\ldots,d'_t,d$ exists. If $d'_i = (U)$ then $\alpha_{d'_i}$ is an axiom of Γ_{TD1} . If we get d'_i from d'_j by rule (1) then we get $\alpha_{d'_i}$ from $\alpha_{d'_j}$ by the first two rules of Γ_{TD1} . If we get d'_i from d'_j and d'_k by rule (2) we get $\alpha_{d'_i}$ from $\alpha_{d'_j}$ and $\alpha_{d'_k}$ by the last two rules of Γ_{TD1} . This implies $\{ \alpha_{d_1},\ldots,\alpha_{d_m} \} \vdash_{\Gamma_{TD1}} \alpha_d$ for any system $d_1,\ldots,d_k$ of join dependencies.

2. Assume that $\{ \alpha_{d_1}, \alpha_{d_2} \} \vdash_{\Gamma_{TD1}} \alpha_d$ and that there is a derivation $\beta_1,\ldots,\beta_t, \alpha_d$ with $d_{\beta_i} \in$ JDEP . If we get β_i (or α_d) from β_j by the first two rules, we get d_{β_i} (or d) from d_{β_j} by the rule (1) . If we get β_i (or α_d) from β_j and β_k by the last rule, we get d_{β_i} (or d) from d_{β_j} and d_{β_k} by the rule (2) . This implies $\{d_1,d_2 \} \vdash_{\Gamma_{TD1}} d$.

This theorem does not state that the system Γ_{JD} is complete for the class JDEP. Theorem 5.3.9. declares that there is no complete Hilbert-type system for the class JDEP. Theorem 5.3.10. shows the axiomatizability by Gentzen-type systems. But theorem 5.3.2. can be applied in several cases for the derivation of new join dependencies. It can be applied especially in the case if there is given a dependency system containing only one join dependency.

Corollary 5.3.3. If the system C of join dependencies is Sheffersch, i.e. generated by one join dependency then from $C \models d$ follows $C \vdash_{\Gamma JD} d$.

There are also other known rules.

In /BEVA 85/ a new rule is presented for TD's. This rule has an analogue in the class JDEP:

(3) $\dfrac{(X_1,\ldots,X_k)\ ,\ (Y_1,\ldots,Y_m)}{(Y_2,\ldots,Y_m)}$ if $MANY((X_1,\ldots,X_k)) \subseteq Y_1$, and $(X_1 \cap Y_1,\ldots,X_k \cap Y_1) \leq (Y_2,\ldots,Y_m)$.

In /BEVA 81/, /THAL 84/ the following rules are introduced:
 i i

$$(4)\quad \frac{\{(X_1,\dots,X_{k_i}) \mid 1\le i\le m\}\ ,\ (Y_1,\dots,Y_m)}{(Z^1_1,\dots,Z^1_{k_1},Z^2_1,\dots,Z^m_{k_m})}$$

for $Z^j_i = \mathrm{MANY}(X^j_i,(X^j_1,\dots,X^j_{k_i})) \cup (X^j_i \cap Y_j)$, $1\le j\le m$, $1\le i\le k_j$.

$$(5)\quad \frac{(X_1,\dots,X_k)\ ,\ \{(Y_1,\dots,Y_m\ ) \mid 1\le i\le k\ \}}{(Z_1,\dots,Z_{\max(m\ )})}$$

for $Z_j = \bigcup_{i=1}^{k} ((X_i \cap Y^i_j\ ) \cup \mathrm{MANY}(Y^i_j,(Y^i_1,\dots,Y^i_{m_i})))$.

The system $\Gamma_{JD''}$ consists of the axiom (A0) and the rules (1) and (3). The system $\Gamma_{JD'''}$ consists of the axiom (A0) and the rules (1) and (4) . The system $\Gamma_{JD^{iv}}$ consists of the axiom (A0) and the rules (1) and (5). The rules (4) and (5) are of practical importance for fast derivations.

Corollary 5.3.4. Let $C \subseteq$ JDEP, $d \in$ JDEP . Then the following statements are equivalent:
(1) $C \vdash_{\Gamma_{JD}} d$. (2) $C \vdash_{\Gamma_{JD''}} d$. (3) $C \vdash_{\Gamma_{JD'''}} d$. (4) $C \vdash_{\Gamma_{JD^{iv}}} d$.

Since the derivations of JD's can be represented using trees with inputs $X_1,..,X_k$ it is useful to restrict the derivation of $d \in \mathrm{JDEP}_k$ from $C \subseteq$ JDEP to derivations of JD's from JDEP_k . Using rule (1), we restrict C to $C \subseteq \mathrm{JDEP}_k$.

Formal system Γ_{JDk} .
Axiom (k0) $(U,U,\dots,U) \in \mathrm{JDEP}_k$.
Rules

(k1) $\dfrac{d}{d'}$ for $d, d' \in \mathrm{JDEP}_k$, $d \le d'$.

(k2) $\dfrac{(X_1,\dots,X_k)\ ,\ (Y_1,\dots,Y_k\ )}{(X_1 \cap Y_1\ ,\ Y_2,\dots,Y_k)}$ if $\mathrm{MANY}((X_1,\dots,X_k)) \subseteq Y_1$, and $X_i \cap Y_1 \subseteq Y_i$, $2\le i\le k$.

We get for $C \subseteq \mathrm{JDEP}_k$, $d \in \mathrm{JDEP}_k$ that $C \vdash_{\Gamma_{JD}} d$ iff $C \vdash_{\Gamma_{JDk}} d$.

There are also other powerful rules for dependencies from JDEP_k , for instance the following for i, j , $1 \leq i,j \leq k$:

$$\text{(k3ji)} \quad \frac{(X_1,\ldots,X_k) \ , \ (Y_1,\ldots,Y_k) \ , \ (Z_1,\ldots,Z_k)}{(V_1,\ldots,V_k)}$$

with

$$V_s = \begin{cases} Z_s \cap Y_s \cap X_s \ \text{u MANY}(Z_s,(Z_1,\ldots,Z_k)) & i=j \ , \ s=i \\ X_s \ \text{u} \ Z_s \cap Y_s \ X_i \ \text{u MANY}(Z_s,(Z_1,\ldots,Z_k)) & s=j, \ s\neq i \\ X_s \ \text{u} \ Y_s \cap X_i \ \text{u MANY}(Y_s,(Y_1,\ldots,Y_k)) \ \text{u} \ Z_s \cap Y_s \cap X_i & s\neq i, \ s\neq j \\ \quad \text{u MANY}(Z_s,(Z_1,\ldots,Z_k)) & \\ Y_s \cap X_s \ \text{u MANY}(Y_s,(Y_1,\ldots,Y_k)) \ \text{u} \ Z_s \cap Y_j \cap X_s \ \text{u} & s=i, \ s\neq j \\ \quad \text{u MANY}(Z_s,(Z_1,\ldots,Z_k)) & . \end{cases}$$

If we want to know the set $C^+ = \{ d \in \ \mid C \mid\!\!\!\text{---}_{\Gamma JD} \ D \}$ then it is sufficient to construct a subset of minimal elements of C^+ , i.e. $C^* = \{ d \in C^+ \ \mid \ d' \in C^+ , d' \leq d \implies d = d' \}$.

In /THAL 84/, an algorithm for construction of C^* is presented. This algorithm uses the d-cover of a set X , $X \subseteq U$ for $d = (Y_1,\ldots,Y_m)$:
$Z(d,X) = (\text{MANY}(Y_1,d) \ \text{u} \ X \cap Y_1 ,\ldots, \text{MANY}(Y_m,d) \ \text{u} \ X \cap Y_m)$.

<u>Example</u>. /SCIO 82/ Let $U = \{A,B,C,E,F,G\}$,
$D = \{(\{A,B,C\},\{B,E,F,G\}), (\{A,B,E\},\{A,C,F,G\}), (\{A,B,C\},\{C,E,F,G\}),$
$(\{A,E\},\{A,F,G\},\{B,F\},\{B,C,G\}) \}$.

Using the d-cover we get the following set

$D^* = \{(\{A,B\},\{B,C\},X_1,X_2) \ \mid \ X_1 \in Z_1 , X_2 \in Z_2 \}$ u
$\{(\{A,B\},\{A,C\},X_1,X_2) \ \mid \ X_1 \in Z_1 , X_2 \in Z_2 \}$ u
$\{(\{A,C\},\{B,C\},X_1,X_2) \ \mid \ X_1 \in Z_1 , X_2 \in Z_3 \}$ u
$\{(\{B,C\},\{A,E\},\{A,F,G\},X_1,X_2) \ \mid \ X_1 \in Z_4 , X_2 \in Z_5 \}$ u
$\{(\{A,C\},\{B,C\},\{A,F,G\},X_1,X_2,X_3) \ \mid \ X_1 \in Z_6 , X_2 \in Z_4 , X_3 \in Z_5 \}$ u
$\{(\{A,B\},\{B,E\},\{C,G\},\{B,F,G\})\}$

where the sets Z_i are defined as follows:

$Z_1 = \{\{A,E\},\{B,E\},\{C,E\}\},$
$Z_2 = \{\{A,F,G\},\{B,F,G\},\{C,F,G\}\},$
$Z_3 = \{\{B,F,G\},\{C,F,G\}\},$
$Z_4 = \{\{B,F\},\{C,F\}\},$
$Z_5 = \{\{B,G\},\{C,G\}\},$

$Z_6 = \{\{B,E\},\{C,E\}\}$.

Furthermore we get $|C^*| = 37$. The 37 JD's can be used for the characterization of all JD's from C^+ , i.e. $d \in C^+$ iff there is some $d' \in C^*$ such that $d' \leq d$.

Now we consider the existence of Armstrong relations.

Theorem 5.3.5. The set JDEP is Armstrong.

Proof /THAL 84/ (see also /GPT 80/).

By $d(r)$ we denote the set $\{ d \in JDEP \mid r \mid\mid\!== d\}$.

Let $C \subseteq JDEP$ be JDEP-closed. Then a set $\underline{R}$ of relations exists with $C = \bigcup_{r \in \underline{R}} d(r)$. If $\underline{R} = \{ r \}$ then r is an Armstrong relation.

Now we prove by induction that one relation r exists for $\underline{R}$ with $C = \{ d \in JDEP \mid r \mid\mid\!==d \}$. Let the existence be proved for $\underline{R'}$ with $|\underline{R'}| = m$.

For $\underline{R}$ with $|\underline{R}| = m+1$ there are two relations r_1, r_2 with

$d(r_1) = \bigcap_{r \in \underline{R}_1} d(r)$, $d(r_2) = \bigcap_{r \in \underline{R}_2} d(r)$, $\underline{R}_1 \cup \underline{R}_2 = \underline{R}$, $|\underline{R}_1| \leq m$, $|\underline{R}_2| \leq m$.

Now a relation r_3 with $d(r_3) = d(r_1) \cap d(r_2)$ will be constructed. Let

$r_1' = \{ ((t_1,1),\ldots,(t_n,1)) \mid (t_1,\ldots,t_n) \in r_1 \}$ and

$r_2' = \{ ((t_1,2),\ldots,(t_n,2)) \mid (t_1,\ldots,t_n) \in r_2 \}$.

1. If C does not contain full crosses then $r_3 = r_1' \cup r_2'$ is a relation with $d(r_3) = d(r_1) \cap d(r_2)$ because of if for $(X_1,\ldots,X_k) \in C$ $t_1,\ldots,t_k$ are elements of r_3 with $t_i(X_i \cap X_j) = t_j(X_i \cap X_j)$ then $t_1,\ldots,t_k$ are either elements of r_1' or elements of r_2' and either in r_1' or in r_2' an element t can be found with $t(X_i) = t_i(X_i)$, $1 \leq i \leq k$.

2. Let C contain full crosses $(X_1^1,\ldots,X_p^1)$, $\ldots,(X_1^m,\ldots,X_1^m)$. Then by theorem 5.2.3. there exist a minimal full cross $(X_1,\ldots,X_k)$, i.e. $\{(X_1,\ldots,X_k)\} \mid= (X_1^i,\ldots,X_s^i)$ $(1 \leq i \leq k)$.

Now let $r_3 = r_3'[X_1] * \ldots * r_3'[X_k]$ for $r_3' = r_1' \cup r_2'$.

Furthermore let $(Y_1,\ldots,Y_l) \in C$ with $\{(X_1,\ldots,X_k)\} \mid\neq (Y_1,\ldots,Y_l)$.

If $r_3 \mid\mid\!=\neq (Y_1,\ldots,Y_l)$ then there are $t_1,\ldots,t_l$ in r_3 such that $t_i(Y_i \cap Y_j) = t_j(Y_i \cap Y_j)$, but there is no t in r_3 with $t(Y_i) = t_i(Y_i)$ $(1 \leq i,j \leq l)$.

Since $r_3[X_i] \mid\mid\!== (Y_1 \cap X_i, Y_2 \cap X_i,\ldots,Y_l \cap X_i)$ for $1 \leq i \leq k$ there are tuples $t_1',\ldots,t_k'$ in r_3 with $t_j'(Y_j \cap X_i) = t_j(Y_j \cap X_i)$ and

$\{t\} = \{t_1'[X_1]\} \times \ldots \times \{t_k'[X_k]\} \subseteq r_3$.

We get $t[Y_j] = t_j[Y_j]$, $1 \leq i \leq l$, in contrary to the assumption

$r_3 \;|\!|\!\neq (Y_1,\dots,Y_l)$. Therefore , $(Y_1,\dots,Y_l) \notin d(r_3)$ for $(Y_1,\dots,Y_l) \in C$.

Using the above presented proof, we get

Corollary 5.3.6. Let $C \subseteq$ JDEP be a JDEP-closed set and $(X_1,\dots,X_k)$ the minimal full cross of C and
$C' = \{ (Y_1,\dots,Y_l) \in C \mid Y_i \subseteq X_j$ or $Y_i \cap X_j = \emptyset$ for any $i,j\}$.
Then $C' \cup \{(X_1,\dots,X_k)\} \models C$.

Now, let $a_D(n)$ denote the maximum size of Armstrong relations for sets C, $C \subseteq$ JDEP (JDEP = JDEP(U) where $U = \{A_1,\dots,A_n\}$) .

Corollary 5.3.7. $a_D(n) > 2^{[n/2]}$.

Proof. For $n = 2k+1$ let $U = \{C,A_1,\dots,A_k,B_1,\dots,B_k\}$. For $n = 2k+2$ let $U = \{C,E,A_1,\dots,A_k,B_1,B_k\}$. Let
$D = \{(\{C,A_i,B_i\},U-\{A_i,B_i\}),(\{A_i,B_i,C\},U-\{C\}) \mid 1 \leq i \leq k \}$, $d = (X_1,X_2,X_3)$ with
$X_1 = \{C,A_1,\dots,A_k\}$, $X_2 = \{C,B_1,\dots,B_k\}$, $X_3 = \{A_1,\dots A_k,B_1,\dots,B_k\}$
($X_1 = \{C,E,A_1,\dots,A_k\}$, $X_2 = \{C,E,B_1,\dots,B_k\}$, $X_3 = \{E,A_1,\dots A_k,B_1,\dots,B_k\}$ for $n = 2k+2$). The JD d is not implied by D (see chapter 3.3). Let r be an Armstrong-relation for D . Let $t_1,t_2,t_3 \in r$ with $t_i(X_i \cap X_j) = t_j(X_i \cap X_j)$ for $1 \leq i < j \leq 3$ such tuples with $\{t\} = \{t_1[X_1]\}*\{t_2[X_2]\}*\{t_3[X_3]\} \not\subseteq r$.
If $t(B_i) = t_1(B_i)$ then $t \notin r$ because of $D \subseteq d(r)$.
If $t(A_i) = t_2(A_i)$ then $t \notin r$, similarly. Similarly, $t(C) \neq t_3(C)$. But then t_1 and t_2 generate with the first group of dependencies in D 2^k tuples. Using t_1 and t_3 we get 2^k tuples from the second group of JD's in C . Thus, r has at least $2^k + 2^k + 1$ tuples.

Using the proof of theorem 5.3.5. we get an important result of independence of schemata /THAL 84/.

Theorem 5.3.8. Let DS = (RS,C) be a database scheme where RS is a relation scheme (U , <u>D</u>, dom) where $U = \{ A_1,\dots,A_n\}$ and C is a set of JD's on U. If there is a full cross $(X_1,\dots,X_k)$ in C then the scheme DS = (RS1,...RSk,C') with RSi = (X_i , <u>D</u>, dom_i) where dom_i is the restriction of dom to X_i and

$$C' = \bigcup_{i=1}^{k} \{ (Y_1 \cap X_i,\dots,Y_l \cap X_i) \mid (Y_1,\dots,Y_l) \in C \}$$

is equivalent to the scheme DS .

This result can be improved only for full hierarchical dependencies using some further dependencies in C' .

In theorem 5.3.2., we have proven that the system Γ_{JD} is JDEP-full. This system can not be extended to a complete system. It is shown in /PETR 89/ that there is a set Σ of join dependencies and a projected join dependency d with $\Sigma \models d$ and with no join dependency d' such that $d' \models d$ and $\Sigma \models d'$. Hence not all inferences of join dependencies consist only of join dependencies. Therefore no modification of the system Γ_{JD} is sufficient. It is claimed in /PETR 89/ that finite axiomatization migth exist for the class $JDEP_k$. This is based on a theorem that arity increase in the derivations is restricted to twice the arity of the initial dependencies. Therefore, the question is still open whether there exists an axiomation of $JDEP_k$. Based on theorem 3.4.4. the following statement is proven in /PETR 89/.

Theorem 5.3.9. There is no finite sound and complete formal system for the class JDEP.

Although the axiomatization of JD's by Hilbert-type systems is impossible, there exists an simple, containing only one rule Gentzen-style formal system which is complete. In Gentzen-type formal systems axioms and rules are of the type
<label>: C ==> d and
<label> : C ==> d
<label> : C ==> d' .
The label is required to guide the derivations, i.e. E: C ==> d is true if $C \models d$. For labels embedded, generalized mutual dependencies (EGMD) are used. Let $E = (X_1,...X_m)$ be an EGMD . The JD $d = (Y_1,..,Y_m)$ is E-based if $X_i \subseteq Y_i$ and $Y_i \cap Y_j \subseteq X_i \cup X_j$ for $1 \leq i < j \leq m$. The following formal system Γ_J uses the rule (5).

Theorem 5.3.10. /BEVE 85/. The formal system Γ_J is sound and complete for JD's.

Formal system Γ_J

Axiom (J0) $E : C \Longrightarrow (X_1,..,X_{i-1},U,X_{i+1},...,X_m)$ for an EGMD $(X_1,...,X_m)$

Rule (J1)

$$\frac{E : C \Longrightarrow d_1,..., E : C \Longrightarrow d_k}{E : C \Longrightarrow (Z_1,...,Z_m)}$$

if $d_i = (Y_1^i,...,Y_m^i)$ are E-based JD's such that for some $(X_1,...,X_m) \in C$

$X_i \cap X_j \cap Y_p^i \subseteq Y_p^j$ for all $1 \leq i,j \leq k$, $1 \leq p \leq m$

and $Z_i = \bigcup_{j=1}^{k} (X_j \cap Y_i^j)$.

6. INCLUSION DEPENDENCIES

The next dependency we will discuss is neither uni-relational nor many-sorted. A great deal of research has gone into understanding single relations, whether they are designed properly. Much less is known about how the relations should fit together. In general, an inclusion dependency (IND) is of the form

$$R\langle A_1, \ldots, A_m\rangle \subseteq S\langle B_1, \ldots, B_m\rangle$$

where R and S are predicates (or relation scheme names), and the A_i's and B_j's are attributes of the corresponding schemes. The inclusion dependency holds for a database if each tuple that is a member of the relation corresponding to the left-hand side is also in the relation corresponding to the right-hand side. Hence, IND's are valuable for database design, since they permit us to selectively define what data must be duplicated in what relations. IND's, together with FD's, are perhaps the most important integrity constraints for relational databases. Although IND's have been extensively utilized for databases, they only recently were subject of theoretical investigations. Their expressive power is not utilized yet. They could , for instance, play a more important role in management of distributed databases (replication).

They also appear when another database scheme, another database model scheme, for instance an entity-relationship scheme, is mapped to the relational model. Yet in another perspective, IND's can be viewed as a relaxation of the controversial universal relation assumption, which requires that all relations in a database be projections of a single universal relation.

IND's are easily to be understood and to be used; they seem to correspond to the way many designers approach their work.

We shall now examine in detail the axiomatization of IND's (chapter 6.1.) and of IND's and FD's together (chapter 6.2.). Further we will study the axiomatization of unary IND's and FD's together which is fundamental for the framework of relational database systems.

6.1. THE CLASS OF INCLUSION DEPENDENCIES

In general, a relational database consists of a set of relations. There are dependencies that associate one relation with another. The inclusion dependency connects the values in a tuple of one relation with the values in a tuple of another relation and are of the form "if some tuple is in this relation, then another tuple must be in that relation". Such constraints represent essentially an operational view of relational design.

<u>Example.</u> Given the following relation schemes RS1 = (U1 , <u>D1</u> , dom1) where U1 = {PARTICIPANT, HOTEL, ADDRESS} and RS2 = (U2 , <u>D2</u> , dom2) where U2 = {LECTURER, LECTURE, TIME}. The databases on DS = (RS1 RS2, C) can be used to represent the information on participants of a conference and lecturers of a conference. It is evident, that the two relations are not independent. Indeed, any lecturer must be a participant of the conference. We can denote this constraint by

R2<LECTURER> $\subseteq$ R1<PARTICIPANT> .

Now we shall introduce inclusion dependencies generally. The <u>weak inclusion dependency</u> (WIND) is a formula

$\forall x1 \ldots \forall xn\ \exists y1 \ldots \exists ym\ (P_1(x1,\ldots,xn) \longrightarrow P_2(y1,\ldots,ym) \wedge xi_1=yj_1 \wedge \ldots \wedge xi_k=yj_k)$

(denoted by $E = E(P_1,i_1,\ldots,i_k;P_2,j_1,\ldots,j_k)$).

For two given relation scheme RS1 = (U1 , <u>D1</u> , dom1) and RS2 = (U2 , <u>D2</u>, dom) where U1 = { A1,...,An} and U2 = {B1,...,Bm} the WIND E can be also denoted by $RS1\langle Ai_1 \ldots Ai_k\rangle \subseteq RS2\langle Bj_1 \ldots Bj_k\rangle$.

If the i_1 are pairwise different and the j_m are pairwise different the WIND E is called <u>inclusion dependency</u> (IND) . If k=1 the WIND E is called <u>unary inclusion dependency</u> (UIND).

WIND's are considered in /MITC 83/, IND's are considered in /CFP 84/.

Now we will introduce another, shorter notion for WIND's. Obviously, for relations r_1 on RS1 and r_2 on RS2, by definition if there is valid in DS= (RS1 RS2, C) the WIND E iff for any tuple t_1 in r_1 there is a tuple t_2 in r_2 such that for any l , $1 \le l \le k$, $t_1(Ai_l) = t_2(Bj_l)$ where A_i and B_j are the corresponding attributes.

Obviously, WIND's are general embedded implicational dependencies and IND's are many-sorted, general embedded implicational dependencies. Equalities can be expressed WIND's with repeated attributes.

We present now the formal system Γ_{IND} of S.H. Lin (/CFP 84/) and prove its completeness and soundness. The proof is taken from /CFP 84/.

Formal system Γ_{IND} .

Axiom (IND0) $R\langle X\rangle \subseteq R\langle X\rangle$ if X is a sequence on U for R about U;

Rules

(IND1) $$\frac{R_1\langle A1,\ldots,Am\rangle \subseteq R_2\langle B1,\ldots,Bm\rangle}{R_1\langle Ai_1,\ldots,Ai_k\rangle \subseteq R_2\langle Bi_1,\ldots,Bi_k\rangle}$$ for each sequence $i_1,\ldots,i_k$ of distinct integers from $\{1,\ldots,m\}$ (permutation and projection)

(IND2) $$\frac{R_1\langle X\rangle \subseteq R_2\langle Y\rangle \quad , \quad R_2\langle Y\rangle \subseteq R_3\langle Z\rangle}{R_1\langle X\rangle \subseteq R_3\langle Z\rangle}$$

__Theorem 6.1.1.__ /CFP 84/ Let C be a set of IND's, and let E be a single IND. The following statements are equivalent:

(1) $C \models E$.

(2) $C \models_{fin} E$.

(3) $C \vdash_{\Gamma IND} E$.

Proof. We shall show that (3) ==> (1) ==> (2) ==> (3) . The system Γ_{IND} is sound. That (1) implies (2) and that (3) implies (1) is immediate. Now we proof (2) ==> (3) .

Assume $C \models_{fin} E$. We must show that $C \vdash_{\Gamma IND} E$.

Let $E = R_1\langle A_1,\ldots,A_m\rangle \subseteq R_2\langle B_1,\ldots,B_m\rangle$, $C = C(R_1,\ldots,R_n)$.

We will inductively create a database $r_1,\ldots,r_n$ for $R_1,\ldots,R_n$, by adding tuples, one at a time.

0.) Let $r_2 = r_3 = \ldots = r_n = \emptyset$ and $r_1 = \{t_1\}$ with

$$t_1(A_i) = \begin{cases} i & \text{if } i \in \{A_1,\ldots,A_m\} \\ 0 & \text{otherwise} . \end{cases}$$

1.) Induction step. Let $R_i(D_1,\ldots D_k) \subseteq R_j(F_1,\ldots,F_k) \in C$, $t \in r_i$ and t'

$$t'(F) = \begin{cases} t(D_i) & \text{if } F = F_i \text{ for some } i , 1 \le i \le k \\ 0 & \text{otherwise.} \end{cases}$$

Then add the tuple t' to r_j , if t' is not in r_j.

Evidently, the resulting database $(r_1,\ldots,r_n)$ on $\underline{D} = \{\{0,1,\ldots,m\}\}$ is finite. It is easy to see that the database also satisfies C, or else the rule 1. could be applied to add another tuple. Since also, by assumption, $C \models_{fin} E$, it follows that the database satisfies E . So, since r_1 contains the tuple t_1 it follows that r_2 contains a tuple t_2 where $t_2(B_i) = i$ $(1 \le i \le m)$

It is sufficient to prove if r_j contains a tuple t with $t(G_m) = i_m \geq 1$ for

$1 \leq s \leq k$ then $C \vdash_{\Gamma_{IND}} R_1\langle A_{i_1},\ldots,A_{i_k}\rangle \subseteq R_j\langle G_1,\ldots,G_k\rangle$.

If $t = t_1$ then the proposition is true since

$C \vdash_{\Gamma_{IND}} R_1\langle A_{i_1},\ldots,A_{i_k}\rangle \subseteq R_1\langle A_{i_1},\ldots,A_{i_k}\rangle$ by (IND0) .

Now we show that the proposition is true about tuple t , under the inductive assumption that it holds for all tuples previously inserted in the database. Assume that the tuple t is inserted in relation r_j as a result of the IND $R_i(D_1,\ldots,D_m) \subseteq R_j(F_1,\ldots,F_m)$ of C and of a tuple t' of r_i . Let us say that the attribute D_v of R_i corresponds to attribute F_v of r_j , for $1 \leq v \leq s$. Let G_q be the attribute of R_i that corresponds to attribute H_q of R_j $(1 \leq q \leq k)$, where the attributes H_q are as in the proposition. Then $t'(G_q) = i_q$, since $t(H_q) = i_q$ $(1 \leq q \leq k)$. Since $t'(G_q) = i_q$, since $t(H_q) = i_q$ $(1 \leq q \leq k)$. Since the IND $R_i(D_1,\ldots,D_m) \subseteq R_j(F_1,\ldots,F_m)$ is in C it follows by (IND1) that

$C \vdash_{\Gamma_{IND}} R_i(G_1,\ldots,G_k) \subseteq R_j(H_1,\ldots,H_k)$.

By inductive assumption the proposition holds when the parts of R_j and t are played by R_i and t' , respectively. Hence

$C \vdash_{\Gamma_{IND}} R_1\langle A_{i_1},\ldots,A_{i_k}\rangle \subseteq R_i\langle G_1,\ldots,G_k\rangle$. So, by (IND2) , it follows that

$C \vdash_{\Gamma_{IND}} R_1\langle A_{i_1},\ldots,A_{i_k}\rangle \subseteq R_j\langle H_1,\ldots,H_k\rangle$, which was to be shown.

In /CFP 84/ it is also shown using the proof that the implication problem for IND's is PSPACE-complete. The finite implication problem for this case is still open. In certain special cases, there is a polynomial-time algorithm for this problem, for example if we confine our attention to IND's of the form $R_1\langle X\rangle \subseteq R_2\langle X\rangle$.

For weak inclusion dependencies, a sound and complete formal system Γ_{WIND} /MITC 83/ is also known.

Formal system Γ_{WIND} .

Axiom (WIND0) $R\langle X\rangle \subseteq R\langle X\rangle$ if X is a sequence of U for R defined on U;

Rules

(WIND1) $$\frac{R_1\langle A_1,\ldots,A_m\rangle \subseteq R_2\langle B_1,\ldots,B_m\rangle}{R_1\langle A_{i_1},\ldots,A_{i_k}\rangle \subseteq R_2\langle B_{i_1},\ldots,B_{i_k}\rangle}$$ for each sequence $i_1,\ldots,i_k$ of integers from $\{1,\ldots,m\}$

(WIND2) $$\frac{R_1\langle X\rangle \subseteq R_2\langle Y\rangle \quad , \quad R_2\langle Y\rangle \subseteq R_3\langle Z\rangle}{R_1\langle X\rangle \subseteq R_3\langle Z\rangle}$$

(WIND3) $$\frac{R_1\langle XY\rangle \subseteq R_2\langle ZZ\rangle \quad , \quad E_1}{E_2}$$ E_2 is obtained from E_1 by substituting X for one or more occurrences of Y

The proof of soundness and completeness of Γ_{WIND} is analogous to 6.1.1. The rule (WIND3) illustrates the additional power of weak inclusion dependencies in comparison with inclusion dependencies.

A WIND $R_1\langle A_1,\dots,A_m\rangle \subseteq R_2\langle B_1,\dots,B_m\rangle$ is <u>typed</u> if $A_i = B_i$ for $1 \le i \le m$.

A set C of WIND's is called <u>acyclic</u> if,

(a) $R_1\langle A_1,\dots,A_m\rangle \subseteq R_1\langle B_1,\dots,B_m\rangle$ in C implies $A_i = B_i$ for $1 \le i \le m$;

(b) There are no distinct predicates $R_1, R_2,\dots, R_n$ ($n>1$) such that C contains $R_1\langle-\rangle \subseteq R_2\langle-\rangle$, $R_2\langle-\rangle \subseteq R_3\langle-\rangle$,..., $R_n\langle-\rangle \subseteq R_1\langle-\rangle$ where - stands for any attributes.

We would like to point out that all the negative complexity results to-date used the power of untyped WIND's to express permutations of the attributes. Using the same power we have the following proposition for acyclic but untyped WIND's, but without using permutations we have also another complexity bound /COKA 83/:

The implication problem for acyclic WIND's alone, is NP-complete.

This proposition can be shown using the formal system Γ_{WIND} or Γ_{IND} and the reducibility of the permutation generation /GAJO 79/ to it.

Now, we consider unary inclusion dependencies and introduce a formal system Γ_{UIND} /KCV 83/.

Formal system Γ_{UIND} .

For all attributes A, B, ...,C

Axiom (UIND0) $R\langle A\rangle \subseteq R\langle A\rangle$

Rules

(UIND1) $$\frac{R_1\langle A\rangle \subseteq R_2\langle B\rangle \quad , \quad R_2\langle B\rangle \subseteq R_3\langle D\rangle}{R_1\langle A\rangle \subseteq R_3\langle C\rangle}$$

From theorem 6.1.1. follows

<u>Corollary 6.1.2.</u> The formal system Γ_{UIND} is sound and complete for implication of UIND's.

In /THAL 84/, nondeterministical inclusion dependencies are introduced. They are of substantial importance for the database design in the entity-relationship approach.

The <u>nondeterministical inclusion dependency</u> (NIND) is a formula

$$\alpha = \forall x_1 \ldots \forall x_n \exists y^1_1 \ldots \exists y^1_{m_1} \ldots \exists y^1_1 \ldots \exists y^1_{m_1} (P_1(x_1,\ldots,x_n) \dashrightarrow$$

$$((P^1(y^1_1,\ldots,y^1_{m_1}) \wedge x_{i_1} = y^1_{j^1_1} \wedge \ldots \wedge x_{i_k} = y^1_{j^1_k}) \vee$$

$$\ldots\ldots\ldots\ldots\ldots\ldots\ldots\ldots\ldots$$

$$\vee (P^1_2(y^1_1,\ldots,y^1_{m_1}) \wedge x_{i_1} = y^1_{j^1_1} \wedge \ldots \wedge x_{i_k} = y^1_{j^1_k})))$$

(denoted by $E = E(P_1,i_1,\ldots,i_k ; P^1_2,j^1_1,\ldots,j^1_k ;\ldots; P^1_2,j^1_1,\ldots,j^1_k)$ or

$E = P_1\langle X\rangle \subseteq \{P^1_2\langle Y_1\rangle,\ldots,P^1_1\langle Y_1\rangle\}$)

where the i_p's , j^1_s's ,..., j^1_t's are pairwise distinct, respectively.

Formal system Γ_{NIND} .

Axiom (NIND0) $P\langle X\rangle \subseteq \{P\langle X\rangle\}$, $P\langle X\rangle \subseteq \{P\langle X\rangle , P\langle X\rangle\}$ for any sequence X on U for P on U ;

Rules

(NIND1)
$$\frac{P_1\langle A_1,\ldots,A_m\rangle \subseteq \{P^1_2\langle B^1_1,\ldots,B^1_m\rangle,\ldots,P^1_2\langle B^1_1,\ldots,B^1_m\rangle\}}{P_1(A_{i_1},\ldots,A_{i_k}) \subseteq \{P^1_2\langle B^1_{i_1},\ldots,B^1_{i_k}\rangle,\ldots, P^1_2\langle B^1_{i_1},\ldots,B^1_{i_k}\rangle\}}$$

(projection and permutation) for each sequence $i_1,\ldots,i_k$ of distinct integers from $\{1,\ldots,m\}$;

(NIND2)
$$\frac{P_1\langle X\rangle \subseteq \{P^1_2\langle Y_1\rangle,\ldots,P^n_2\langle Y_n\rangle\} , \{P^i_2\langle Y_i\rangle \subseteq \{P_3^{i1}\langle Z_{i1}\rangle,\ldots,P_3^{ik_i}(Z_{ik_i})\} \mid 1\le i\le n\}}{P_1\langle X\rangle \subseteq \{P_3^{11}\langle Z_{11}\rangle ,\ldots, P_3^{1k_1}\langle Z_{1k_1}\rangle ,\ldots, P_3^{nk_n}\langle Z_{nk_n}\rangle \}}$$

(transitivity)

(NIND3)
$$\frac{P_1\langle X\rangle \subseteq \{ P^1_2\langle Y_1\rangle,\ldots,P^n_2\langle Y_n\rangle \}}{P_1\langle X\rangle \subseteq \{ P^1_2\langle Y_1\rangle,\ldots,P^n_2\langle Y_n\rangle,P_3\langle Z\rangle\}}$$ for a sequence Z on P_3 of the same length as X

The proof of theorem 6.1.1. can be used to prove the soundness and completeness Γ_{NIND} for implications of NIND's .

We remark that in /CAVI 83/ another different kind of dependencies the so-called exclusion dependency was introduced and considered. This class of dependencies can be understood as the strongest a-inclusion dependencies. In general, an <u>exclusion dependency</u> is a sentence of the form

$F \; = \; R\langle A_1,\ldots,A_m\rangle \; || \; S\langle B_1,\ldots,B_m\rangle$
where R and S are predicates (relation names) and the A_i's and B_j's are attributes of R and S , respectively.
Given the relation schemes R = (U1 , D1, dom1) and S = (U2 , D2, dom2) where $U1 = \{A_1,\ldots,A_m\}$, $U2 = \{B_1,\ldots,B_m\}$ and
$dom1(A_1) \times\ldots\times dom1(A_m) = dom2(B_1) \times\ldots\times dom2(B_m)$.
The exclusion dependency F holds for a (R , S , C)-database (r_1,r_2) if
$r_1[\{A_1,\ldots,A_m\}] \cap r_2[\{B_1,\ldots,B_m\}] = \emptyset$.

6.2. INCLUSION DEPENDENCIES AND THEIR INTERACTION WITH FUNCTIONAL DEPENDENCIES

Functional and inclusion dependencies are the most important and fundamental database integrity constraints, and they are mainly used in all data models. Recently, their interaction has been investigated in several papers (/CFP 84/, /CHVA 83/, /COKA 83/, /KCV 83/, /MITC 83/, /SCOR 82/). These interrelation constraints are of importance even in connection with functional dependencies. For instance, we are given the relation schemes RS1 = (U1 , D1 , dom1), RS2 = (U2 , D2 , dom2) where $U1 = \{A_1,\ldots,A_p\}$, $U2 = \{A_{p+1},\ldots,A_n\}$ and the FD's RS1 : $\{A_1,\ldots,A_s\} \dashrightarrow \{A_1,\ldots,A_p\}$, RS2 : $\{A_{p+1},\ldots,A_t\} \dashrightarrow \{A_{p+1},\ldots,A_n\}$ are in C .
In /KOBA 85/, the inclusion dependency $RS1\langle A'_1,\ldots,A'_k\rangle \subseteq RS2\langle A_{p+1},\ldots,A_{p+k}\rangle$ is called <u>onto constraint</u> for $k \geq t-p$,
and the inclusion dependency $RS1\langle A_1,\ldots,A_k\rangle \subseteq RS2\langle A'_1,\ldots,A'_k\rangle$ is called <u>into constraint</u> for $k \geq s$, and $\{A_1,\ldots,A_k\}$ is called <u>foreign key</u> in RS2 .
The existence of an into (onto) constraint implies the existence of an into (onto) correspondence between the relations. If $k = t-p$ then the correspondence is many-to-one. If $k < t-p$ then it becomes many-to-many. Therefore, it is possible to define for schemes (RS1,...,RSk , C) relationship constraints as a set $\{ R_0\langle X\rangle \subseteq R_i\langle Y_i\rangle \mid 1\leq i\leq k\}$ if every Y_i is a key of R_i . The notion of relationship constraints bridges the gap between the relation model and the entity-relationship model.

In order to utilize dependencies in the database design process one must be able to test it for logical implication, i.e. does a set of dependencies logically imply another dependency. It is known, when only functional dependencies are given or when only inclusion dependencies are given, the implication problem is decidable and an axiomatization exists. Now we discover that things get more complicated when both kinds of dependencies are put together. The first disappointing observation is that implication and finite implication do not coincide for the union of these classes.

Theorem 6.2.1. /CFP84/ There is a set C of FD's and UIND's and a single UIND E such that $C \models_{fin} E$ but $C \not\models E$.

Proof. Let $C = \{\{A\} \rightarrow \{B\},\ R<A> \subseteq R<B>\}$ and E be $R<B> \subseteq R<A>$. First we show that $C \not\models E$ using the relation $r = \{(i+1,i) \mid i \geq 0\}$. It is obvious that $r \models C$ but $r \not\models E$.
Now let r be a finite relation satisfying C . We now show that r satisfies E, that is , $r[\{B\}] \subseteq r[\{A\}]$. Since $r \models C$ it follows $|r[\{B\}]| \leq |r[\{A\}]|$ and $|r[\{A\}]| \leq |r[\{B\}]|$. But since $r[\{A\}] \subseteq r[\{B\}]$ and since $r[\{A\}]$ and $r[\{B\}]$ are finite, then we have $r[\{A\}] = r[\{B\}]$, so $r[\{B\}] \subseteq r[\{A\}]$. This was to be shown.

Implication for generalized functional and inclusion dependencies have an unusual property. Remember, a dependency F follows from a set C of dependencies by k-ary implication if there is some subset of k dependencies from C that implies F . A formal system $\Gamma = (Ax, Ru)$ is k-ary if each rule in Ru is at most k-ary.

Theorem 6.2.2./CFP 84/ For no k there exists a k-ary complete axiomatization for IND's and FD's. For no k there exists a k-ary complete axiomatization for finite implication of FD's and IND's. There is no finite axiomatization for (finite) implication of FD's and IND's.

Sketch of the proof. Let k and n be two fixed natural numbers such that $k<n$, P, Q_0 , S_n be 3-ary relation schemes (i.e. $R = (U,\underline{D},dom)$ with $|U| = 3$)
$P = (\{A,B,C\},\underline{D},dom1)$, $Q_0 = (\{A,B,C\},\underline{D},dom2)$, $S_n = (\{B,C,D\},\underline{D},dom3)$ and G_i $(1\leq i\leq n)$ and S_i $(0\leq i<n)$ 2-ary relation schemes , i.e. $G_i = (\{B,C\},\underline{D},dom4)$,
$S_i = (\{B,C\},\underline{D},dom4)$. Define C as the set of dependencies
$\{ P<A,B> \subseteq Q_0<A,B>\ ,\ P<B,C> \subseteq S_n<B,D>\ ,\ Q_0:\{A\} \rightarrow \{C\},\ S_n:\{C\} \rightarrow \{D\}\} \ \cup$
$\{ P<B> \subseteq G_i<B> \mid 1\leq i\leq n \} \ \cup$
$\{ P<B> \subseteq S_i<B>\ ,\ S_i<B,C> \subseteq G_{i+1}<B,C> \mid 0\leq i<n \} \ \cup$
$\{ S_i<B,C> \subseteq G_i<B,C>\ ,\ G_i:\{B\} \rightarrow \{C\} \mid 0\leq i\leq n\}$.

Define F as $P:\{A\}\to\{C\}$. (Remember that if P' is a relation scheme on $U = \{A_1,\ldots,A_n\}$ and if $X,Y \subseteq U$ then we call $P':X\to Y$ functional dependency of P'.)

Now we can show $C \models F$ and $F \notin C^{+k}$ for the k-ary closure C^{+k} of C under $\models$.

For the finite implication we can define the following C and F and prove the same: Let $P_i = (\{A,B\},\underline{D},doml)$ for $0 \leq i \leq k$ be relation schemes and
$C = \{ P_i:\{A\}\to\{B\}\ ,\ P_i<A> \subseteq P_{(i+1) \bmod k}<B> \mid 0 \leq i \leq k\}\ ,\ F = P_k<A> \subseteq P_0<B>$.

In this proof sets of IND's are used which are not acyclic. In /SCOR 82/ it is proved that for restricted sets of IND's and FD's the models defined by this sets and the models defined by the universal relational approach are equivalent to each another in power.

A set C of IND's is <u>confluent</u> if whenever the IND's $P<A> \subseteq S<B>$ and $P<A> \subseteq T<E>$ are implied by C there exists a scheme P' such that the IND's $S<B> \subseteq P'<D>$ and $T<E> \subseteq P'<D>$ are also implied by C .

A set C of IND's is <u>key-invariant</u> if for all IND's $P<X> \subseteq P'<Y>$ in C , Y is a key of P' .

A set C of IND's is <u>union-invariant</u> if whenever the IND's $P<X> \subseteq S<Y>$ and $P<W> \subseteq S<Z>$ are implied by C then so is $P<WX> \subseteq S<YZ>$.

A set C of IND's is <u>effluent</u> if whenever the IND's $T<A> \subseteq P<D>$ and $S<B> \subseteq P<D>$ for attributes A, B, D are implied by C then for all Q such that there are sequences of IND's in C with $Q<X_0> \subseteq Q_1<Y_1>\ ,\ Q_1<X_1> \subseteq Q_2<Y_2>\ ,\ldots,$
$Q_k<X_k> \subseteq T<Y_{k+1}>$ and $Q<X'_0> \subseteq Q'_1<Y'_1>\ ,\ Q'_1<X'_1> \subseteq Q'_2<Y'_2>\ ,\ldots,$
$Q'_k<X'_k> \subseteq S<Y'_{k+1}>$ there exists an attribute E such that $Q<E> \subseteq T<A>$ and $Q<E> \subseteq S<B>$ are implied by C .

Now, the databases defined by sets of FD's and IND's which are effluent, acyclic, key-invariant, union-invariant and confluent and the databases defined by the universal relational approach are equivalent to each another in power. E. Sciore argued that these restrictions should hold in any well-designed relational database scheme.

If we permit inclusion dependencies we can assume that an attribute (metavariable) appears only once in a database scheme; that is, if an attribute A is in U for the relation scheme $RS = (U\ ,\ \underline{D},\ dom)$ where $U = \{ A_1,\ldots,A_n\}$ then it is in no other scheme. This restriction simplifies the notion of sets of

IND's and FD's if we use sequences of attributes instead of sets for the notion of FD's.

The paper /MITC 83/ presents a formal system $\Gamma_{VIND,FD}$ that is complete for general, but not for finite implication of VIND's and FD's. The rules differ from those of the system Γ_{IND} and $\Gamma_{1,FD}$. One inference rule (WF 33) yields dependencies which mention attributes that are not used in the hypotheses.

Formal system $\Gamma_{VIND,FD}$.

Axioms

(WF 01) $XY \longrightarrow Y$ for sequences X,Y of attributes which appear in the same relation scheme;

(WF 02) $X \subseteq X$ for sequences X of attributes which occur in the same predicate ;

Rules

(WF 11) $$\frac{X \longrightarrow Y}{XW \longrightarrow YZ}$$ when all attributes in the sequence Z appear in W

(WF12) $$\frac{X \longrightarrow Y \;,\; Y \longrightarrow Z}{X \longrightarrow Z}$$ transitivity

(WF 13) $$\frac{X \longrightarrow Y}{W \longrightarrow V}$$ where W and V list precisely the same attributes as X and Y , respectively (permutation, redundancy)

(WF 21) $$\frac{A_1,\ldots,A_n \subseteq B_1,\ldots,B_n}{A_{i_1},\ldots A_{i_k} \subseteq B_{i_1},\ldots,B_{i_k}}$$ where $1 \le i_j \le n$ for all j (permutation, projection, redundancy)

(WF 22) $$\frac{X \subseteq Y \;,\; Y \subseteq Z}{X \subseteq Z}$$ (transitivity)

(WF 23) $$\frac{A,B \subseteq C,C \quad , \quad E}{E'}$$ where E' is obtained from E by substituting A for one or more occurrences of B (substitution)

(WF 31) $$\frac{WV \subseteq XY \;,\; X \longrightarrow Y}{W \longrightarrow V}$$ where $|X| = |V|$ (pullback)

(WF 32) $$\frac{UV \subseteq XY \;,\; UW \subseteq XZ \;,\; X \longrightarrow Y}{UVW \subseteq XYZ}$$ where $|X| = |U|$ (collection)

$$\text{(WF 33)}\quad \frac{U \subseteq V \ , \ V \rightarrow \{B\}}{U,A \subseteq V,,B}$$ where A is an attribute which in the same scheme as U (attribute introduction)

A WIND $A_1,\ldots,A_k \subseteq B_1,\ldots,B_k$ is said to be <u>m-ary</u> if $k \leq m$.
Now we show that (finite) implication of FD's and WIND's is reducible to (finite) implication of FD's and binary WIND's.

<u>Theorem 6.2.3.</u> /CHVA 83/ Let C be a finite set of FD's and WIND's, and let E be a FD (resp. a WIND). Then we can effectively construct a finite set C' of FD's and binary WIND's and a FD (resp. a unary IND) E' such that $C \models E$ iff $C' \models E'$ and $C \models_{fin} E$ iff $C' \models_{fin} E'$.

Construction of the proof. W.l.o.g., let all the WIND's in $C \cup \{E\}$ be m-ary and not (m+1)-ary. We denote a sequence $A_1,\ldots,A_m$ of attributes by $\underline{A}$. We can view a sequence as a list of elements. When we enclose the sequence in parentheses, e.g. $(\underline{A})$, we refer to it as an element in the domain of sequences. The proof is based on a grouping mechanism. WIND's can be represented by equivalent grouped binary WIND's.
We construct a set C" of FD's and WIND's. We introduce new attributes $H_1,\ldots,H_m,H$. Now

$$C'' = \{\ \underline{H} \rightarrow H \ ,\ X \rightarrow \underline{H}\ \} \cup$$
$$\{\ A_i,(\underline{A}) \subseteq H_i,H \ ,\ B_i,(\underline{B}) \subseteq \underline{X},X \ \mid\ \underline{A} \subseteq \underline{B} \in C \cup \{E\}\ ,\ 1 \leq i \leq m\ \}.$$

If E is an FD then $E' = E$. If E is the WIND $\underline{A} \subseteq \underline{B}$ then we define E' as the UIND $(\underline{A}) \subseteq (\underline{B})$.

In /MITC 83/ and /CHVA 83/ it is shown that the implication and the finite implication problem for functional dependencies and weak inclusion dependencies are recursively unsolvable. Therefore, the implication and the finite implication problems for FD's and binary WIND's are undecidable. In the proof it is pointed out that functional dependencies force projections of a relation to be functions, and weak inclusion dependencies can express equality between compositions of functions. This reduces the word problem for monoids and finite monoids to the implication and finite implication problem for dependencies. Since implications for finite monoids /BØRG 85/ are not recursively enumerable, there is no complete, recursively enumerable axiomatization for finite database implication.

For restricted sets of WIND's we get the following property from the formal system $\Gamma_{WIND,FD}$.

Corollary 6.2.4. The implication problem for acyclic sets of WIND's and FD's is decidable in exponential space.

An analogous result is shown in /COKA 83/ for restricted sets of typed WIND's. The implication problem for acyclic sets of typed WIND's and sets of FD's is NP-hard /COKA 83/. This directly follows from the reduction from 3-SAT /GAJO 79/.

Another restriction is the class of full inclusion dependencies , e.g. dependencies of the form

$$.(P(x_1,...,x_n) \text{ ---> } P'(x_{i_1},...,x_{i_k}))$$

The implication problem and the finite implication problem for sets of full inclusion dependencies and of functional dependencies are the same, and therefore decidable. This proposition follows directly from corollary 3.1.1.

In the literature, two kinds of domain dependencies are introduced and uniquely named. We distinguish between these kinds. The first domain dependency can be understood as a special unary inclusion dependency, the second as a special general functional dependency.

Given a relation scheme RS = (U , D, dom) where U = { $A_1,...,A_n$} . This scheme can be also understood as an extension of a scheme of a "real world" relational database by a special domain relation. For two relation schemes RS1 = (U1 , D1, dom1) , RS2 = (U2 , D2, dom2) where U1 = {$A_1,...,A_n$}, U2 = {$B_1,...,B_m$} and a subset X of U1 with a length m , the general domain dependency IN(RS1(X), RS2(U2)) means that the X-entry in each tuple of relations r_1 on RS1 must be a member of the set r_2 on RS2. Therefore, general domain dependencies can be understood as full inclusion dependencies.

These domain dependencies are applied in Codd's /CODD 79/ Extended Relational Model in which there is made a distinction between different relation schemes in a database scheme. Relations can perform a subordinate role in describing relations of some other type (characteristic or property relations). They can perform a superordinate role in interrelating relation of other relation schemes (associative relations). If they perform neither of the above roles, they should be considered as kernel relations. A tuple may not appear in a property relation unless its key appears in the corresponding kernel relation. A tuple can exist in an associative relation if the tuples it interrelates also exist in the kernel relations.

Some people claim that in practice, we encounter only WIND's that have a single attribute on each side of the containment. Theorem 6.2.1. shows that the class of UIND's and FD's is a class for which implication and finite implication problems are not equivalent, but both problems are, nevertheless and as a refreshing surprise, solvable. The (finite) implication problem for the class of UIND's and FD's is reducible to the (finite) satisfiability problem for a decidable class of formulas /DRGO 79/.

For axiomatization of the class of UIND's and FD's, we consider the interaction between UIND's and FD's. There is, indeed, more evidence that UIND's interact with other dependencies in a simpler fashion, than general WIND's do. There is an interaction between these subclasses because FD's can force a column to be finite by forcing it to be a singleton set. Specifically, if a relation r satisfies $\emptyset \dashrightarrow \{A\}$, then $|r[\{A\}]| = 1$. Thus, for example, $\emptyset \dashrightarrow \{A\}$ and $R\langle B\rangle \subseteq Q\langle A\rangle$ imply $\emptyset \dashrightarrow \{B\}$ and $Q\langle A\rangle \subseteq R\langle B\rangle$. A result of /KCV 83/ is that this example is the only way in which FD's and UIND's interact. Moreover, the formal systems Γ_{UIND} and Γ_{GEID} together with this interaction form a complete and sound formal system for general embedded implicational dependencies and unary inclusion dependencies. In /KCV 83/ a sound and complete axiomatization for finite implication of FD's and UIND's is presented. The cycle rules of $\Gamma_{FD,UIND}$ are in fact unsound for infinite structures /CFP 84/. Now we present the sound and complete formal system $\Gamma_{FD,UIND}$ of /KCV 83/ .

Formal system $\Gamma_{FD,UIND}$.

Axioms

(FD,UIND 01) $R : XY \dashrightarrow Y$ for sets X,Y of attributes which appear in the same scheme R

(FD,UIND 02) $R\langle A\rangle \subseteq R\langle A\rangle$

Rules

(FD,UIND 1) $$\frac{R : X \to Y \ , \ R : Y \to Z}{R : XV \dashrightarrow ZV}$$ for sets X,Y,Z,V of attributes which appear in the same scheme R (extended transitivity)

(FD,UIND 2) $$\frac{R\langle A\rangle \subseteq Q\langle B\rangle \ , \ Q\langle B\rangle \subseteq T\langle C\rangle}{R\langle A\rangle \subseteq T\langle C\rangle}$$ (transitivity)

For every odd positive integer k :

$$\text{(FD,UIND 3k)}\quad \frac{\begin{array}{ll} R_0 : \{A_0\} \dashrightarrow \{A_1\}, & R_0\langle A_1\rangle \subseteq R_2\langle A_2\rangle, \\ R_2 : \{A_2\} \dashrightarrow \{A_3\}, & R_2\langle A_3\rangle \subseteq R_4\langle A_4\rangle, \\ \ldots\ldots\ldots\ldots & \\ R_{k-1} : \{A_{k-1}\} \dashrightarrow \{A_k\}, & R_{k-1}\langle A_k\rangle \subseteq R_0\langle A_0\rangle \end{array}}{\begin{array}{ll} R_0 : \{A_1\} \dashrightarrow \{A_0\}, & R_2\langle A_2\rangle \subseteq R_0\langle A_1\rangle, \\ R_2 : \{A_3\} \dashrightarrow \{A_2\}, & R_4\langle A_4\rangle \subseteq R_2\langle A_3\rangle, \\ \ldots\ldots\ldots\ldots & \\ R_{k-1} : \{A_k\} \dashrightarrow \{A_{k-1}\}, & R_0\langle A_0\rangle \subseteq R_{k-1}\langle A_k\rangle \end{array}} .$$

The class of FD's and UIND's is one of the smallest known classes containing FD's and for which no Armstrong relation exists /KCV 83/.

7. DEPENDENCIES IN RELATIONS WITH NULL VALUES AND INCOMPLETE INFORMATIONS

In many database applications, the knowledge of the real world modeled by the database is incomplete. A lot of research has been devoted to the problem of querying these so-called incomplete databases. In any real-world database, there will be entries having values that are "special", in the sense that they are not drawn from the value set for that entry. Some of such special values are of the meaning "value unknown", "item inapplicable", "value exists but cannot be stored", "value is not complete classified" etc. (14 different types of null values are well known /ANSI 75/).

This chapter presents some of the problems that arise from the assumption that null values exist in some relational database. Here, the terms null value and incomplete information are primarily used in the meaning of "a value exists but is unknown" (Chapter 7.1), "a value exists in some subset of a domain set" (Chapter 7.2) and "a value is at present unknown but connected with another value by semantics of the database" (Chapter 7.3).

In chapter 7.4, one assumption of database theory is rejected. There are reasons for this rejection. When a database is created, it is not always possible to have complete information about the data. In the relational model, lacking data are usually represented as "null values". Grant/GRAN 79/ introduced two kinds of null values: The first to represent the fact that the corresponding attribute value does not exist and the second to represent the fact that the corresponding attribute value exist, but that the value is not known. In practical cases, even more kinds of null values are often necessary to be handled. There are several types of incomplete information as follows:

(A) Null values :

(A1) A value exists or not.

- (A 1.1) It is known that a value does not exist.
- (A 1.2) It is not known whether or not a value exists.
- (A 1.3) It is known that a value exists but this value is unknown.

(A2) A set of values exist, but only an upper bound for the (maximal) cardinality of the set or only some part of the set is known.

(B) Partial information of values.

(B1) Some part of a value does not exist or is unknown.

(B2) Some part of a value is known and means a set of corresponding values.

Database systems usually require the users to specify values for all fields of the records. However, frequently some values are unknown, which means that we have to introduce the concept of information incompleteness from a theoretical viewpoint. We will focus our attention in chapter 7.4 on only one major area: null values in keys.

An important rule for relational databases seems to be that, for integrity reasons, information about an unidentified (or inadequately identified) object is never recorded in these database (too sharp a contrast with non-relational databases). Thus, the primary key attribute of each base relation is not permitted to include null values of either type. But, with respect to the real world, the database can be incomplete in the sense that not all facts needed and corresponding to the state of the real world are stored in the database. This is possible for all components of a record. This kind of normal incompleteness stems from our restricted knowledge of the real world. As our knowledge of the real world changes the database will have to be adjusted. The database is adjusted to the real world by inserting, deleting and modifying records, i.e. by performing updates on the database. This is everyday computer processing practice which normally does not raise any semantic problems. One should however be careful about the assumptions made for modeling of the real world. One of these common assumptions is the convention on forbidden null values in primary keys: None of the attributes of the primary key may ever obtain an undefined, unknown value, since otherwise we would not know what entity a tuple with an undefined value of the primary key represents. This assumption is a very useful one for searching a record and other practical purposes. This assumption is not necessary. In /KATY 79/ this assumption is rejected since this assumption does not allow compound attributes as long as such compound values are units of updates. The modeling of data dependencies with compound attributes becomes difficult. Therefore the restriction of the nonexistence of null values is relaxed in /KATY 79/ as follows: No primary key value x of any tuple does not coincide with one of any other tuple even if the null value in x is replaced by possible values appearing in those attributes. It is proved in /KATY 79/ that if r = r' + r" for a relation defined on X + Y + Z where the set of X-values of r' contains no null value and any X-value of r" is a concatenation of null values, then r can be obtained by forming the OR-join of r(X+Y) and r(X+Z) providing X ->->Y¦Z holds in r' and a non-existence dependency from X to Y or from X to Z holds in r" . A non-existence dependency from X to Y means that if the set of X-values consists of only null values, then the set of Y-value also consists of null values only. This approach is extended. The only requirement is that the tuples should be distinguishable.

Given a relation scheme $RS = (U, \underline{D}, dom)$ where $U = \{A_1, \ldots, A_n\}$. An **extended tuple** on RS is a function $t : U \longrightarrow \bigcup_{D \in \underline{D}} Pow(D)$ with $t(A) \subseteq dom(A)$ for $A \in U$. If there is defined an order on U ($U = \{A_1, A_2, \ldots A_n\}$) then the extended tuple can be represented by $(t(A_1), \ldots, t(A_n))$.

For singleton sets $t(A)$, the parentheses {} can be omitted.

We denote by $T_-(RS)$ the set of all extended tuples on RS.

Any subset r of $T_-(RS)$ is called **extended relation** (on RS) (or relation if only those are considered).

Given a sequence $DRS = RS_1, \ldots, RS_m$ of compatible relation schemes where $RS_i = (Ui, \underline{Di}, domi)$, $1 \le i \le m$.

By an **incomplete DRS-database** a database $M = (r_1, \ldots, r_m)$ of extended relations r_i on RS_i is understood.

If for each i, $1 \le i \le k$, each $A \in Ui$, for each t from r_i the set $t(A)$ is singleton or empty the incomplete DRS-database M is called **database with null values**. If $t(A) = \emptyset$ then we write also $t(A) = -$ (for unknown).

If for each i, $1 \le i \le k$, each $A \in Ui$, for each t from r_i the set $t(A)$ is non-empty the incomplete RS-Database M is called **database with incomplete information**.

W.l.o.g., we now deal only with uni-relational incomplete database $M = (r)$.

Example 7.1. Consider an accident ward. For each actual accident victim the hospital management is interested in the room number, the name, the address, they are living, the kind of injury and the arrival time. We can represent this information in a table called patient.

ROOM	NAME	ADDRESS	INJURY	TIME
1	Müller	-	cardiac infarct	sunday, 16
-	-	-	skull fracture	monday, 19
2	Maier	Dresden	-	monday, 20
1	Müller	Pirna	leg fracture	sunday, 16

A relation scheme that can be used for this purpose is

PATIENT $= (U, \underline{D}, dom)$ with

U = { ROOM, NAME, ADDRESS, INJURY, TIME },

$\underline{D}$ = { set-of-room-numbers, set-of-last names, set-of-towns, set-of-injuries, set-of-days-and-times }, and

the function dom is obvious.

But for the case of this accident ward, there are known also different integrity constraints as, e.g.,

- no room has more than 5 beds,
- rooms 2, 3 have only one bed, each.

7.1. DATABASES WITH NULL VALUES

D. Maier /MAI 83/ introduces disjunctive existence constraints for the purpose of specifying where "missing value" null may appear in a relational database.

For a set $X, Y_1, \dots, Y_m$ of subsets of U, a <u>disjunctive existence constraint</u> (DEC) has the form $X \Longrightarrow \{Y_1, Y_2, \dots, Y_m\}$.

Given a tuple t of a relation r of a database with null values M = (r). If X is a subset of U, then if for each attribute A in X t(A) is non-empty, we write t(X)!.

A database with null values M = (r) satisfies a disjunctive existence constraint $X \Longrightarrow \{Y_1, \dots, Y_m\}$ iff for each t in r t(X)! implies that there is an i, $1 \le i \le m$, such that $t(Y_i)$!. (Denoted by $M \,||\!\!=\, X \Longrightarrow \{Y_1, \dots, Y_m\}$).

A database satisfies a set of disjunctive existence constraints if the database satisfies every disjunctive existence constraint in this set.

There is an axiomatization of disjunctive existence constraints using its equivalence with monotone functional dependencies.

We are given a database with null values M = (r), a disjunctive existence constraint $X \Longrightarrow \{Y_1, \dots, Y_m\}$ and a monotone functional dependency $X \dashrightarrow Y_1 \dots Y_m$.

Let $r = \{t_1, \dots, t_k\}$.

We define $r' = \{t_1, \dots, t_{2k}\}$ and $M' = (r')$ as follows

$$t_i(A) = i \quad \text{for all } A$$

$$t_{i+k}(A) = \begin{cases} i & \text{if } t_i(A) = 0 \\ i+k & \text{if } t_i(A) \neq 0 \end{cases} \quad \text{for any } A \in U,\ 1 \le i \le k .$$

We get that $r \mid\mid\!\!== X ==> \{Y_1,\ldots,Y_m\}$ iff $r' \mid\mid\!\!== X --> Y_1 \vee\ldots\vee Y_m$.

Now we are given a database $M = (r)$, a disjunctive existence constraint $X ==> \{Y_1,\ldots,Y_m\}$ and a monotone functional dependency $X --> Y_1 \vee\ldots\vee Y_m$.

Let $r = \{t_1,\ldots t_k\}$. We define a database with null values $M' = (r')$ as follows:

$r' = \{t_{ij} \mid 1 \le i \le k,\ 1 \le j \le k\}$

$$t_{ij}(A) = \begin{cases} t_i(A) & \text{if } t_i(A) = t_j(A) \\ 0 & \text{if } t_i(A) \ne t_j(A) \end{cases} \qquad \text{for any } A \in U .$$

We get that $r \mid\mid\!\!== X --> Y_1 \vee\ldots\vee Y_m$ iff $r' \mid\mid\!\!== X ==> \{Y_1, \ldots, Y_m\}$.

We now define the formal system Γ_{DEC} .

Formal system Γ_{DEC} .

Axioms (DEC 0) $XY => \{X\}$ for $X,Y \subseteq U$.

Rules. For $X,Y_1, \ldots,Z_1,\ldots,Z_{ij}, \ldots \subseteq U$

$$\text{(DEC 1)} \quad \frac{X==> \{Y_1,\ldots,Y_m\}}{X==> \{Y_1,\ldots,Y_m,Z\}} \qquad \text{(augmentation)}$$

$$\text{(DEC 2)} \quad \frac{X==>\{Y_1,\ldots,Y_m\},\ X==>\{Z_1,\ldots,Z_k\}}{X==>\{Y_1Z_1,\ldots,Y_1Z_k,Y_2Z_1,\ldots,Y_mZ_m\}} \qquad \text{(union)}$$

$$\text{(DEC 3)} \quad \frac{X==>\{Y_1,\ldots,Y_m\},\ \{Y_i==>\{Z_{i1},\ldots,Z_{i\,k(i)}\} \mid 1 \le i \le m\}}{X==> \{Z_{11},\ldots,Z_{1\,k(1)},Z_{21},\ldots,Z_{m\,k(m)}\}} \qquad \text{(transitivity)}$$

Using the completeness theorem for monotone functional dependencies and the above constructed equivalence, we get

__Theorem 7.1.1__. The formal system Γ_{DEC} is sound and complete for the class of disjunctive existence constraints.

In /GOLD81/ another proof of this theorem is given.

Now, functional dependencies will be examined in the light of databases with null values. Four notions of validity of FD will be introduced and considered. Another less sharper approach to validity of FD's is given in /VASS 80/.

We define two equivalence relations $=_X$, $\sim_X$ for subsets X of the attributed set U .

We are given a relation scheme RS = (U , _D_, dom) where U = $\{A_1,\ldots,A_n\}$, relation r on RS and a subset X of U .

Two tuples t,t' from r are equivalent with respect to X (denoted by $t =_X t'$) if t(X) = t'(X), t(X)! and t'(X)!.

Two tuples t,t' from r are weak equivalent with respect to X (denoted by $t \sim_X t'$) if for any A $\in$ X following conditions hold t(A)!, t'(A)!, t(A) = t'(A) or one of the following conditions is false: t(A)!, t'(A)! .

Now there are four approaches to define the validity of a functional dependency X --> Y in r:

1. for all t,t' $\in$ r from $t =_X t'$ follows $t =_Y t'$
   (denoted by r ¦¦== X --> Y);
2. for all t,t' $\in$ r from $t =_X t'$ follows $t \sim_Y t'$
   (denoted by r ^{1}¦¦== X --> Y);
3. for all t,t' $\in$ r from $t \sim_X t'$ follows $t =_Y t'$
   (denoted by r ^{2}¦¦== X--> Y);
4. for all t,t' $\in$ r from $t \sim_X t'$ follows $t \sim_Y t'$
   (denoted by r ^{3}¦¦== X --> Y).

The last validity can be understood as a condition that a completion of M exist in which X --> Y is valid.

<u>Corollary 7.1.2</u>.

1. If r ^{2}¦¦== X --> Y then r ^{3}¦¦== X --> Y and r ¦¦== X --> Y .
2. If r ¦¦== X --> Y then r ^{1}¦¦== X --> Y .
3. If r ^{3}¦¦== X --> Y then r ^{1}¦¦== X --> Y .
4. The inversion of 1., 2., 3. does not hold.
5. It does not hold that from r ¦¦== X --> Y follows r ^{3}¦¦== X --> Y
   or from r ^{3}¦¦== X --> Y follows r ¦¦== X --> Y .

The axiomatization of the implication of these four approaches is different.

<u>Corollary 7.1.3</u>.

1. If r ¦¦== X --> Y and r ¦¦== Y --> Z then r ¦¦== X --> Z .
   If r ¦¦== X --> YZ then r ¦¦== X --> Y .
   If r ¦¦== X --> Y and r ¦¦== X --> Z then r ¦¦== X --> YZ .
   It holds r ¦¦== XY --> Y .

2. If r ^{1}||== X --> YZ then r^{1}||== X --> Y .
   If r ^{1}||== X --> Y and r ^{1}||== X --> Z then r ^{1}||== X --> YZ .
   If r ^{1}||== X --> Y then r ^{1}||== XZ --> YZ .
   It holds r ^{1}||== XY --> Y .
   In general, from r^{1}||== X --> Y and r^{1}||== Y --> Z the condition r ^{1}||== X --> Z does not follow.
3. If r ^{2}||== X --> Y and r ^{2}||== Y --> Z then r ^{2}||== X --< Z .
   If r ^{2}||== X --> YZ then r ^{2}||== X --> Y .
   It does not hold r ^{2}||== X Y --> Y in general.
4. If r ^{3}||== X --> Y and r ^{3}||== Y --> Z then r ^{3}||== X --> Z .
   If r ^{3}||== X --> YZ then r ^{3}||== X --> Y .
   If r ^{3}||== X --> Y and r ^{3}||== X --> Z then r ^{3}||== X --> YZ.
   It holds r ^{3}||== X Y --> Y .
5. Armstrong's formal system Γ_{FD} is sound for functional dependencies defined on databases with null and the requirement of the ||==-validity or the ^{3}||==-validity.
6. The rules defined by 2. above form a sound and complete set of inference rules and axiom for the ^{1}||==-implication of functional dependencies.

For the proof of the last part we can repeat the proof of chapter 4.2. In /ATMO 84/ is presented an extension of the rules for the case of presence of DEC's. Given a relation scheme RS = (U , D , dom) and the DEC d = Ø==>{U'} for some subset U' of U.
For a relation r on (RS,{d}) the following rule is valid for X, Y, Z $\subseteq$ U with Y - X $\subseteq$ U' :
If r ^{1}||== X --> Y and r ^{1}||== Y --> Z then r ^{1}||== X --> Z (null-transitivity).
It is shown that for the scheme (RS,d) the rules presented in corollary 7.1.3 part 2 and the null transitivity-rule form a complete and sound formal system.

From corollary 7.1.3 follows that whether for implication defined by ^{1}||== nor for implication defined by ^{2}||== , a representation of implications with Boolean functions cannot exist.

The most important characterization of implications of different kinds of functional dependencies is the characterization in the world of 2-tuple relations or databases with 2-tuple relations.

For a set C of FD's , the FD $X \dashrightarrow Y$, a relation scheme RS = (U , $\underline{D}$, dom) , the set $\underline{R}_{RS,0}$ of RS-databases with null values we define that:

from C follows strong $X \dashrightarrow Y$ if for any $r \in \underline{R}_{RS,0}$ it holds $r\ ||{\neq}\ C$ or $r\ ||{=}{=}\ X \dashrightarrow Y$;

from C follows 1 -weak $X \dashrightarrow Y$ if for any $r \in \underline{R}_{RS,0}$ it holds $r\ {}^{1}||{=}{=}\ C$ or $r\ {}^{1}||{=}{=}\ X \dashrightarrow Y$,

from C follows 2-weak $X \dashrightarrow Y$ if for any $r \in \underline{R}_{RS,0}$ it holds $r\ {}^{2}||{=}{=}\ C$ or $r\ {}^{2}||{=}{=}\ X \dashrightarrow Y$,and,

from C follows 3-weak $X \dashrightarrow Y$ if for any $r \in \underline{R}_{RS,0}$ it holds $r\ {}^{3}||{=}{=}\ C$ or $r\ {}^{3}||{=}{=}\ X \dashrightarrow Y$.

Using the chase method and the database $r_1 = \{t_1, t_2\}$ (for ${}^{1}||{=}{=}$, ${}^{2}||{=}{=}$) and $r_2 = \{t_1, t_3)$ (for $||{=}{=}$, ${}^{3}||{=}{=}$) for a given FD $X \dashrightarrow Y$ with

$t_1(A) = a$ for $A \in U$

$$t_2(A) = \begin{cases} 0 & A \in X \\ b & A \in Y \\ a & A \notin XY \end{cases}$$

$$t_3(A) = \begin{cases} a & A \notin Y - X \\ b & A \in Y - X \end{cases}$$ we get

<u>Corollary 7.1.4</u>. Suppose that the rule
Ru: from $\{d_1, \ldots, d_m\}$ follows strong (1-weak, 2-weak, 3-weak) d is not sound for d, $d_1, \ldots, d_m$. Then, there is a 2-tuple database with null-values for which $\{d_1, \ldots, d_m\}$ holds but d does not hold in the corresponding notion.

Using the different validities, we can say that X is a <u>sure key</u> in r if $r\ {}^{2}||{=}{=}\ X \dashrightarrow U$ and for each $t \in r$, t(X)! and that X is a <u>possible key</u> in r if $r\ {}^{1}||{=}{=}\ X \dashrightarrow U$.

In practice, there will usually be restrictions where nulls should appear in a relation. For instance, nulls are forbidden in any component of the primary key of a relation. Therefore, normally sure keys are candidates for primary keys.

Applying the formal definition of equality, a lot of different problems arises for multivalued and binary join dependency.

We are given a database $M = (r)$ with null values and a partition $\{X,Y,Z\}$ of U.

In /LIEN 79/ we find the following definition:

The binary join dependency (XY, XZ) holds in r iff whenever two tuples t,t' with $t(X)!$, $t'(X)!$ and $t(X) = t'(X)$ are in r so is also a tuple t'' with $t(XY) = t''(XY)$ and $t'(XZ) = t''(XZ)$ (denoted by $r \models\!\models (XY, XZ)$).

We can define the following formal system Γ_{NBJ}/LIEN 79/:

Formal system Γ_{NBJ}.

Axioms (NBJ 0) $(U,\emptyset)$

Rules for $d_1 = (X_1,X_2)$, $d_2 = (Y_1,Y_2) \in JDEP_2$

(NBJ 1) $\dfrac{d_1}{d_2}$ $\quad d_1 \leq d_2$

(NBJ 2) $\dfrac{(X_1,X_2)\ ,\ (Y_1,Y_2)}{(X_1 \cap Y_1\ ,\ X_2Y_2)}$ $\quad X_1 \cap X_2 = Y_1 \cap Y_2$.

Corollary 7.1.5. The formal system Γ_{NBJ} is sound for the class of binary join dependencies on databases with null values.

The following statements shows the difference between databases and databases with null values.

Corollary 7.1.6. The following rules are not sound for binary join dependencies on databases with null values:

(1) $\dfrac{(X_1,X_2)\ ,\ (Y_1,Y_2)}{(X_1 \cap (X_2Y_1)\ ,\ X_2Y_2)}$ $\quad (X_1,X_2),(Y_1,Y_2) \in JDEP_2$;

(2) $\dfrac{(X_1,X_2)\ ,\ (Y_1,Y_2)}{(X_1 \cap Y_1\ ,\ Y_2)}$ $\quad (X_1,X_2),(Y_1,Y_2) \in JDEP_2$ with $X_1 \cap X_2 \subseteq Y_1 \cap Y_2$ and $X_2 \subseteq Y_2$.

Using a database $r = \{t_1,t_2,t_3\}$ with $t_1(X_1 \cap Y_1 \cap Y_2)!$, $t_2(X_1 \cap Y_1 \cap Y_2)!$, $t_3(X_1 \cap Y_1 \cap Y_2)!$, $t_3(X_1) = t_1(X_1)$, $t_3(X_2) = t_2(X_2)$ and in which not holds that $t_1(X_1 \cap Y_1 - Y_2)!$ and $t_2(Y_2-(X_1 \cap Y_1))!$ we get that a database (r) with null values exists with the properties $r \models\!\models (X_1,X_2)$, $r \models\!\models (Y_1,Y_2)$ and $r \not\models\!\models (X_1 \cap Y_1,Y_2)$.

Theorem 7.1.7. The formal system Γ_{NBJ} is sound and complete for binary join dependencies without full crosses on databases with null values.

Proof. Suppose C is a set of binary join dependencies not being binary full crosses and (XV, XZ) cannot be derived in Γ_{NBJ} from C. Remember that for $X \subseteq U$ the partition $(W_1,\ldots,W_m)$ of $U-X$ is called dependency basis for (X,C) if a dependency (XV',XZ') can be derived from C in Γ_{NBJ} iff

$$V' = \bigcup_{i, W \cap V' \neq \emptyset} W_i .$$

Let $(W_1,\ldots,W_m)$ be the dependency basis for (X,C).

Now we will construct a database with null values (r) with $r \models C$, $r \not\models (XV,XZ)$ and $r = \{t_1,\ldots,t_{2m}\}$. Let for $1 \leq i \leq m$, $A \in U$,

$$t_{2i-1}(A) = \begin{cases} i & \text{if } A \in X \\ 0 & \text{if } A \in W_i \\ 0 & \text{if } A \notin W_i X \end{cases} \quad \text{and}$$

$$t_{2i}(A) = \begin{cases} i & \text{if } A \in X \\ 1 & \text{if } A \in W_i \\ 0 & \text{if } A \notin W_i X \end{cases}$$

If for $(S,T) \in C$ and for $i \neq j$ $S \cap T \cap W_i \neq \emptyset$ and $S \cap T \cap W_j \neq \emptyset$ then (S,T) holds trivially in r because of for any $t \in r$ we refute $t(S \cap T)!$.

If for $(S,T) \in C$ and some i $S \cap T \subseteq XW_i$ we get $r \models (S,T)$ using the definition of the dependency basis for (X,C). Finally, it is required to show that (XV,XZ) does not hold in r. There must be a j such that $W_j \cap V \neq \emptyset$ and $W_j \cap V \neq W_j$. Therefore, $r \not\models (X(V \cap W_j), U-(V \cap W_j))$. Since $(XW_j, U-W_j)$ holds in r, by soundness of Γ_{NBJ} (XV,XZ) must not hold in r.

7.2. DATABASES WITH INCOMPLETE INFORMATION

Although most of the databases in use are databases by definition of chapter 1, indefiniteness can occur as a result of incomplete knowledge about the real world. For example, we could know that the blood type of John is a or b, but insufficient is available to determine exactly which blood type John has. This fact could be represented by the tuple (John, {a,b}) of the relation BLOOD-TYPE.

If we extend our notion of databases to databases with incomplete information, a lot of problems arises in connection to the definition of relational operations, to the dealing with negative information in databases and to dependency theory.

In usual databases, negative information is implicitly represented. A negative information $-P_1(x_1,...,x_n)$ is assumed to be true if we fail to prove $P_1(x_1,...,x_n)$ from the existing set of tuples in relation r_1 of the database. This representation is called "Closed World Assumption" by Reiter /REIT 78/. The closed world assumption is logically equivalent to adding a new component $M^- = (r_1^-,...,r_k^-)$ to the database $M = (r_1,...,r_k)$ where $r_1^- = T(RS) - r_1$. This approach is not applicable for databases with incomplete information. This is shown by the example BLOOD-TYPE above mentioned.

Now we are given a (uni-relational)(n-ary) database (r) with incomplete information where $r \subseteq Pow^+(dom(A_1)) x ... x Pow^+(dom(A_n))$ where by $Pow^+(G)$ is denoted the set of all non-empty subsets of G. We say that a tuple t of r is completely classified with respect to $A \in U$ if $t(A)$ is singleton. A tuple t of R is <u>completely classified</u> with respect to $X \subseteq U$ if for any $A \in X$ it is completely classified with respect to A.

Let us state that in the extreme case when all tuples are completely classified with respect to U, the system (r) coincides with the database defined in section 1 (i.e. is a database without incomplete information).

Given two databases with incomplete information $M_1 = (r_1)$, $M_2 = (r_2)$. We say that M_2 is a <u>refinement</u> of M_1 (denoted by $M_1 \leq M_2$) if for any tuple $t_1 \in r_1$ there is one tuple $t_2 \in r_2$ such that for each $A \in U$ it holds $t_2(A) \subseteq t_1(A)$ and if for any $t_2 \in r_2$ there is a tuple $t_1 \in r_1$ such that for each $A \in U$ it holds $t_2(A) \subseteq t_1(A)$.

A database $M^+ = (r^+)$ is called a (minimal) completion of (r) if all tuples of r^+ are completely classified with respect to U and if $M \leq M^+$ (and if no proper database (r') , $r' \subset r^+$ is a refinement of (r)).

Now we are able to define the <u>generalized closed world assumption</u>. A DRS-formula $- P_1(c_1,...,c_n)$ can be assumed to be true in M if and only if $P_1(c_1,...,c_n)$ is not true in any minimal completion of M.

A DRS-formula $- P_1(x_1,...,x_n)$ can be assumed to be satisfiable in M if and only if $P_1(x_1,...,x_n)$ is unsatisfiable in some minimal completion of M.

<u>Corollary 7.2.1</u>. A database M is a database without incomplete information if and only if it has exactly one minimal completion.

<u>Corollary 7.2.2</u>. 1. Let M = (r) be a database, $k_A = \max_{t \in r} |t(A)|$ for $A \in U$

and $k_M = \prod_{A \in U} k_A$. Then k_M is an upper bound on the number of minimal completions of M.

2. For any set $\{k_A \mid A \in U, k_A \in N\}$ a database with incomplete information $(\underline{D}_U,R)$ exists which has $\prod_{A \in U} k_A$ different minimal completions.

It is not quite obvious how to generalize the meaning of $r \mid\mid== \alpha$ to the case of databases with incomplete information. It seems that, basically speaking, two different approaches to the problem are possible. The first approach to interpret formulas in M is to refer them to a completion. The second approach of interpreting formulas in a database with incomplete information is to assume that the meaning of $P(x_1,\ldots,x_n)$ is: "it is known that $P(x_1,\ldots,x_n)$ is satisfied in reality". In other words, the interpretation of a formula $\alpha(x_1,\ldots,x_n)$ in database M coincides with the usual interpretation of $\alpha(x_1,\ldots,x_n)$ in a completion of M. Since these two approaches are equivalent we now define for a database M = (r) with incomplete information and for the corresponding scheme RS and language L(RS):

$r \mid\mid== []\alpha$ iff for every completion r' of r $r' \mid\mid== \alpha$;

$r \mid\mid== \langle\rangle\alpha$ iff there is a completion r' of r with $r' \mid\mid== \alpha$.

We have introduced an additional unary semantical connective [] to our language L(RS). By an extended formula, any formula which (possibly) contains [] and <> is designated. This languages will be denoted by L(RS).

The idea of introducing the modal connective [] to the language was suggested by the Kripke models for the modal logic 84 /LIPS 81/.

Using the equivalence of $\langle\rangle\alpha$ and $-[]-\alpha$ we get the following important fact for dependency theory.

<u>Corollary 7.2.3</u>. $r \mid\mid== [](\alpha \rightarrow \beta)$ iff $r \mid\mid== \langle\rangle\alpha \rightarrow []\beta$.

This fact can be used, for instance, for the definition of validity of functional dependencies in databases with incomplete information.

Given two tuples t,t' of r , $X \subseteq U$ and a Boolean function f.

We say that t is <u>sure</u> (<u>possible</u>) <u>equivalent to t' with respect to X</u> (denoted by $[]t =_X t'$ ($\langle\rangle t =_X t'$)) if in every (some) completion of r it holds t(X) = t'(X). Similar $[]\ t =_f t'$ and $\langle\rangle\ t =_f t'$ are defined for the function f.

The database M = (r) surely satisfies (f,g) (denoted by (r)¦¦== [] (f,g)) if M' ¦¦== (f,g) for any completion M' of M . We get the following

Theorem 7.2.4. Let (f,g) be a generalized functional dependency and M = (r) be a database with incomplete information. Then M ¦¦== [](f,g) iff for any t,t' $\in$ R from <>t $=_f$t' follows [] t $=_g$t' .

7.3. CONTEXT-DEPENDENT NULL VALUES

Up to now, null values have deterministic meanings and they are represented by a bounded number of null symbols in databases, for instance ahead with only one null value - or Ø. The null value "at present unknown" indicates the case that this attribute is defined for this object but we do not know its real value. Often, especially in a large database or in a database derived from another by universal relation approach, null values occur in a database and have different meanings. Therefore, we lose information applying the approach of chapter 7.1. The corollary 7.1.6 demonstrates the limitations of this approach. Now we will introduce another viewpoint on null values with better possibilities to obtain information. We observe that in this approach, such problems with negative influence do not exist. Context-dependent null values are defined by the "local" context of the database and are first examined in /NCHT 87/. Possible equivalent null values are identified with respect to the relation, i.e. to the context.

We are given relation scheme RS = (U , D, dom) with $U = \{A_1,...,A_n\}$. A relation scheme RS_Φ = (U ,$D\Phi$, dom) with null-value set $\Phi = \{\Phi_1,...,\Phi_n\}$ is given by extension of the domain sets $dom(A_i)$ by the infinite null value sets Φ_i . A relation r can be defined in an analog approach.
A tuple t on RS_Φ can be defined as a function f with the domain U and the property $t(A_i) \in dom(A_i) \cup \Phi_i$. A relation r on RS_Φ is then a finite set of tuples on RS_Φ .

For sets $X \subseteq U$ and a RS_Φ-database r three binary relations can be introduced as follows:
Two tuples t,t' from r are said to be X-equivalent (denoted by $t \approx_X t'$) if for any $A \in X$ $t(A) \notin dom(A_i)$ and $t'(A) \notin dom(A_i)$ or $t(A) = t'(A)$. Let be $\delta,\delta' \in \Phi_i$ for some $A_i \in U$. The two null values δ,δ' are said to be (r,X)-equivalent (denoted by $\delta \approx_{R,X} \delta'$) if for any t and t' in r with $t(A) = \delta$ and

$t'(A) = \delta'$ there exist null values $\delta_0, \ldots, \delta_m$ in Σ, tuples $t_0, t_1, \ldots, t_{2m+1}$ in r such that $\delta_0 = \delta$, $\delta_m = \delta'$, $t_0 = t$, $t_{2m+1} = t'$ and

$t_i \approx_{X-\{A\}} t_{i+1}$ for $0 \leq i \leq 2m$,

$t_{2j}(A) = t_{2j+2}(A) = \delta_j$ for $0 \leq j < m$ and

$t_{2j-1}(A) = t_{2j+1}(A) = \delta_j$ for $1 \leq j \leq m$.

Intuitively, $\approx_{r,X}$ means that the null values have the same context or have the same meaning in r(X) at present.

Obviously, $\approx_X$ and $\approx_{R,X}$ are equivalence relations.

Prior to definition of validity of binary join dependencies in M a third equivalence relation is required.

Two tuples t,t' from r are said to be X-weak Y-equivalent in r (denoted by r $t \approx_{X,Y} t'$) for $Y \subseteq X \subseteq U$ if for any $A \in Y$

either $t(A) = t'(A)$ or $t(A) \approx_{r,X} t'(A)$.

We are given a partition X,Y,Z of U.

The binary join dependency (X Y, X Z) holds weakly in M = (r) (denoted by r ||==* (X Y, X Z)) if for any two tuples t,t' from r with $t \approx_{XY,X} t'$ there exist two tuples t", t"' in r such that

$$t''(A) = \begin{cases} r(A) & \text{if } A \in XY \\ r'(A) & \text{if } A \in Z \end{cases}$$

$$t'''(A) = \begin{cases} r(A) & \text{if } A \in XZ \\ r'(A) & \text{if } A \in Y \end{cases} .$$

For a set of dependencies $C \subseteq JDEP_2$ and a RS-database M with null value-sets M ||==*C holds iff for any binary JD $(X,Y) \in C$ M ||==* (X,Y).

From a set C of binary join dependencies follows weakly a binary join dependency (X,Y) (denoted by C ||==* (X,Y)) if M ||==*(X,Y) holds for any RS-database with null-value sets M with M ||==* C .

In /NCHT 87/ is proven the soundness and completeness of the following formal system for weak implication.

Formal system $\Gamma_{BJD,W}$.

Axiom. (U,U)

Rules. For binary JD $d_1 = (X_1,X_2)$, $d_2 = (X'_1,X'_2)$

(W1) $\dfrac{d_1}{d_2}$ $d_1 \leq d_2$

(W2) $\dfrac{(X_1,X_2)\ ,\ (X'_1,X'_2)}{(X_1 \cap (X_2X'_1)\ ,\ X_2X'_2)}$.

This system is similar to the system $\Gamma_{JD2''}$. On the other hand, the system Γ_{NBJ} is similar to the system $\Gamma_{JD2}v$ which is known to be incomplete for binary join dependencies.

7.4. KEY SETS IN RELATIONS WITH NULL VALUES

A key functionally determines all the attributes of the relation and is used to distinguish the tuples of a relation. For relations with null values the concept of distinguishability can be introduced instead of the more strong concept of keys on fully defined attributes.

Let K be a set of non-empty subsets of U and C_K the following function of integrity constraints: For any relation r on RS

$C_K(r)$ = 1 iff for any different tuples t, t' from r there exists a set Y in K such that t(Y)! , t'(Y)! and t(Y) ≠ t'(Y) .

If r ¦¦== C_K then it will be also denoted by r ¦¦== K .
The set K will be called the **key set** of r .

Example 7.1. Recall example 7.1. The set
{{ ROOM }, { NAME }, {ADDRESS}, {INJURY}, {TIME}}
is a key set of the relation presented in example 1. Another key set would be the set K' = {{ROOM, TIME}, {NAME, TIME}, {ADDRESS, TIME}, {INJURY, TIME}} . Obviously there is no one-element key set of PATIENT . For the presented relation r it holds also
r ¦¦== {{INJURY},{TIME}} which is a typical key set for the usual way of communicating in accident wards. It is not valid that r ¦= {{INJURY,TIME}} ,i.e. r ¦¦=≠ {{INJURY,TIME}} .

As we have already seen, a key set may be considered as a set of candidates for possible keys. When we tackle the problem of which key sets are of importance, it is useful to split the problem. In this chapter we consider the problem in dependence on one relation. There are as a minimum two approaches for keys in relational databases with null values:
1. The assumption on forbidden null values in (primary) keys, i.e. , only one-element key sets are taken into consideration. This is the usual point of view. But this approach may be too restrictive (see example 1).

2. The assumption of key set existence or distinguishability, i.e., key sets which consist of one-element elements are taken into consideration. It is this point of view which, in practice, matters.

Between these two approaches lie many other approaches which allow us to describe more precisely the keys we desire. The database system itself finds the best presentation for keys.

Let $RS = (U,\underline{D},dom)$ be a relation scheme, r a relation on RS and K a key set of r, i.e. $r \mid\mid\!= K$. The set K is said to be **nonredundant** w.r.t. r iff it holds $r \mid\!\neq K - \{Y\}$ for any $Y \in K$.

The following fact enables a reduction algorithm for key sets of relations to be set up.

<u>Corollary 7.4.1.</u> If K is a nonredundant key set of r and there are sets Y, Z in K with $Y \subseteq Z$ then $\{K - \{Z\}\}\{Z-Y\}$ is also a nonredundant key set of r.

Using corollary 7.4.1. a key set of r can be easily constructed because of the fact that for any non-empty subset $X = \{B_1,...,B_m\}$ of U, a relation $r = \{t,t_1,...,t_m\}$ such that $\{X\}$ is a key set of r and no proper subset of X forms a key. Therefore this property is non-trivial. An example of a relation with is:

$t(A) = 0$ for $A \in X$, $t(A) = -$ for $A \in U - X$;

$t_i(A) = 1$ for $A = B_i$, $t_i(A) = 0$ for $A \in U - \{B_i\}$ $(1 \leq i \leq m)$.

But also for sets of relations on RS corollary 7.4.1 is valid.

A nonredundant w.r.t. r key set K which is a Sperner-set, i.e. for $Y,Z \in K$ none of the properties $Y \subseteq Z$, or $Z \subseteq Y$ holds, is called <u>reduced key set</u>.

Let us denote by $Fak(n)$ the number

$$\binom{n}{\lfloor \frac{n}{2} \rfloor}.$$

<u>**Theorem 7.4.2**</u>. There is a relation scheme $RS = (U,\underline{D},dom)$ with $|U| = n$ such that for every k, $1 \leq k \leq Fak(n)$ there exists a relation r on RS which has a reduced key set with k elements.

Proof. W.l.o.g. we prove the theorem only for $k = Fak(n)$. We construct a relation r with a key set K with k elements for

$K = \{ X \subseteq U \mid |X| = [n/2] \}$.

The first tuple consists of nothing but 1's. The other tuples can be grouped in blocks for each possible variant of representing [n/2] attributes. Each block contains for the corresponding variant in this [n/2] - 1 entries 1's and the remaining entries are i's excluding one of the n - [n/2] + 1 remaining attributes for each element of the block in which attribute the tuple has a null value - .

For $n = 4$, see the relation below:

1	1	1	1
1	2	2	-
1	3	-	3
1	-	4	4
5	1	5	-
6	1	-	6
-	1	7	7
8	8	1	-
9	-	1	9
-	10	1	10
11	11	-	1
12	-	12	1
-	13	13	1

If we choose [n/2] places in a tuple then we find there are either only 1's or at least one number different from i. Therefore the tuple t_i is uniquely determined. Any $X \subseteq U$ with $|X| = [n/2]$ is an element of the key set. It is easy to see that no set $X \subseteq U$ with $|X| < [n/2]$ can be an element of the key set. Therefore, a nonredundant key set is a Sperner-set.

Given a set system K. A set system K' is called a **refinement** of K if for any $Y \in K$ there are $Z_1, \ldots, Z_k \in K'$ such that

$Y = Z_1 \ldots Z_k$.

By $\cup K$ we denote the union of all elements of K.

Corollary 7.4.3. If K is a key set of r then any refinement of K is also a key set of r. If K is key set of r, K' a refinement of K and $K'' \subseteq K'$ a nonredundant key set of r then $\{ Y \cap (\cup K'') \mid Y \in K \}$ is also a key set of r.

Corollary 7.4.4. If K is a key set of r then there exists a nonredundant key set $K' = \{X_1,\ldots,X_k\}$ with $|X_i| = 1$ for $1 \le i \le k$ and $\cup K' \subseteq \cup K$.

A nonredundant key set $K = \{X_1,\ldots,X_k\}$ of r with $|X_i| = 1$ for $1 \le i \le k$ is called a **minimal key set**.

Minimal key sets are useful for the solution of algorithmic problems however normally a key set should express moreover also an information about the appearance of null values in tuples. Therefore using only minimal key sets we are loosing information. Nevertheless, for minimal key sets, using methods in /DEME 79/ we have

<u>Theorem 7.4.5</u>. The largest number of minimal key sets that can occur in any relation r on $RS = (U,\underline{D},dom)$ with $|U| = n$ is Fak(n) . There is a relation scheme $RS = (U,\underline{D},dom)$ such that for every k , $1 \le k \le$ Fak(n) , there exists a relation r on RS with minimal key sets with k elements.

Proof. It is obvious that two distinct minimal key sets K, K' of r cannot contain each other. Therefore the set of all minimal key sets is a Sperner-set. The first part of the theorem now follows immediately from Sperner's theorem /SPER28/.

We will now construct a relation r with m = Fak(n) minimal key sets. The first tuple of r consists of nothing but 1's. The other tuples contain [n/2]-1 1's in all possible ways while the remaining entries of the i-th tuple are i's ($2 \le i \le m$) . Obviously, if we choose [n/2] attributes in an extended tuple then we find either only 1's or at least one number i different from 1 . Then the tuple t_i is uniquely determined. Any X with $|X| =$ [n/2] is a key and therefore $K_X = \{\{A\} \mid A \in X\}$ is a minimal key set. It is easy to see that no set K with $|K| <$ [n/2] can be a minimal key set.
Using the same construction the stronger statement 2 of the theorem can be proved analogously.

This result enables us to use all known algorithms and propositions on keys in relational structures without null values. But the minimal key set is only the minimal limit for the existence of key properties in a fixed relation. A key set of a relation which is not minimal comprises, as already noticed, also other useful information on the occurrence of null values in distinct attributes. An analogous approach would be the simultaneous consideration of minimal key sets and disjunctive existence constraints /THAL'87/ together. It can be of importance to use the maximal information on the occurrence of null values in tuples from a given relation. For the solution of this problem we have to use redundant key sets. We introduce two notions for a given scheme RS = (U,D,dom), a relation r on RS and tuples t,t' from r :

$Def(t,t') = \{ A \in U \mid t(A) \neq - , t'(A) \neq - \}$,

$Diff(t,t') = \{ A \in U \mid t(A) \neq - , t'(A) \neq -, t(A) \neq t'(A) \}$,

$Def(r) = \{Def(t,t') \mid t,t' \in r , t \neq t' \}$,

$Diff(r) = \{Diff(t,t') \mid t,t' \in r , t \neq t' \}$.

__Corollary 7.4.6.__ The sets Def(r) and Diff(r) are key sets of r iff $\emptyset \notin Diff(r)$.

Sets Def(r) and Diff(r) satisfying $\emptyset \notin Diff(r)$ can be considered as the "maximal" key sets. They contain the maximal available information on null values in the relation r . Therefore the size of these sets is important.

A relation r on RS is called **normal** if $\emptyset \notin Diff(r)$.
Using the proof method of theorem 4.4.7 we get

__Theorem 7.4.7.__. The largest size of Diff(r) in any normal relation r on RS = (U,D,dom) with $|U| = n$ is $2^n - 1$. For any k , $1 \leq k \leq 2^n - 1$, there exists a relation r on RS with $|Diff(r)| = k$.

This property clearly shows that such a notion of maximality is useless for practical problems.

The __maximal key set__ M of r is the maximal subset $K_{max}(r)$ of Diff(r) with the property $X \in K_{max}(r)$ & $Y \in Diff(R)$ & $Y \subseteq X \longrightarrow X = Y$, i.e. M is the set of all minimal elements of Diff(r) .

Obviously, $K_{max}(r)$ is a Sperner-set.

__Theorem 7.4.8__. To every Sperner system $K \subseteq \{X \mid X \subseteq U \}$ a relation r on RS = (U,D,dom) can be constructed with the maximal key set K .

For the proof we use the relation presented in the proof of theorem 7.4.5 and the methods presented in the proof of theorem 7.4.2. Therefore , the proof can be omitted.

Using theorem 7.4.2 and theorem 4.4.9 we get now an estimation for the number of all maximal key sets K of relations r on RS = (U,D,dom) with $|U| = n$. Let $m = Fak(n)$ $u = \ln(n)/\sqrt{n}$, $v = 1/2^n$, then there are constants c and c' such that there exist at least $2^{(1 + c\, u)m}$ and at most $2^{(1 + c'\, v)m}$ different maximal key sets.

Another property of a key set M is the irreducibility of elements, i.e.. the minimality w.r.t. the number of necessary attributes in every element of M.

The key set K of r is called irreducible w.r.t. r iff for any $Y \in K$, $Y' \subseteq Y$, $Y' \neq Y$ the set $(K - \{Y\})\{Y'\}$ is not a key set of r.

Corollary 7.4.9 . If K is an irreducible key set w.r.t. r then $K \subseteq$ Diff(r).

8. HORIZONTAL DECOMPOSITION DEPENDENCIES

In the study of the relational database model, the vertical decomposition of relations into projections of these relations was emphasized since the introduction in /CODD 72/. The use of vertical decompositions always requires some constraints to be satisfied, for instance a join dependency or a functional dependency, in order to be able to regain the original relation by taking the join of its projections. In /ARDE 80/, /THAL 84/ and AABM 80/ the idea of D. Smith and J. Smith /SMSM 77/, to decompose a relation horizontally into restrictions of these relations, using the union as composition operator, was formalized, using Codd-functional and multivalued dependencies. Such horizontal decompositions /DBPA 83/ are useful in the normalization of schemata in which hidden constraints are involved.

Horizontal decompositions are especially useful to treat exceptions to constraints /DBPA 82/. In this chapter, we aim at to characterize conceptual relations among schemata obtained by horizontal decomposition, the properties of a special class of dependencies and introduce a new class of union constraints. Although the papers in relation to horizontal decomposition are in minority, the horizontal decomposition theory is of same importance as the vertical decomposition theory. This horizontal decomposition theory is especially useful for databases which must represent "real world" situations, in which there always are exceptions to rather severe constraints like functional dependencies and multivalued dependencies.

8.1. THE HORIZONTAL DECOMPOSITION

It is well known that functional and multivalued dependencies are the favorite constraints used to decompose relation schemata. This privilege is surely due to the simplicity of the concept of these dependencies, and to their widespread appearance in the real world. However, in a great number of applications it is required to allow violation of some FD's, i.c. FD's that are desired but that do not hold in the whole relation.

Initially, we consider a pair of schemes (RS,C) and (DRS',C') and a pair of languages L(RS) and L(DRS') where $RS = (U,\underline{D},dom)$, $DRS' = RS_1,\ldots,RS_m$, $RS_i = (U,\underline{U},dom)$, $U = \{A_1,\ldots,A_n\}$, $1 \leq i \leq m$ and C' is a set of formulas over RS_i in which P_j for $j \neq i$ does not occur, i.e. $C' = C_1C_2\ldots C_m$.

Now, the inclusion and equivalence of the schemata can be characterized.

<u>Theorem 8.1.1</u>. /AABM 80/ (1) If for any i, $1 \leq i \leq m$, it holds that $C_i \models C$ then $(DRS', C') \leq (RS,C)$.

(2) If for some i, $1 \leq i \leq m$, it holds that $C \models C_i$ then $(RS,C) \lesssim (DRS',C')$.

(3) If for any i, $1 \leq i \leq m$, it holds that $C_i \models C$ and for some j, $1 \leq j \leq m$, it holds that $C \models C_j$ then (RS,C) is weakly equivalent to (DRS',C').

(4) The scheme (RS,C) is equivalent to (DRS',C') if the following conditions are satisfied:

(i) $C_i \models C$ for any i, $1 \leq i \leq m$;

(ii) $C \models C_j$ for some j, $1 \leq j \leq m$;

(iii) $\models -(C_i) \vee -(C_j)$ for any i,j, $1 \leq i < j \leq m$,

where $-(C_k) = -d_{k1} \vee -d_{k2} \vee \dots \vee -d_{k\,t(k)}$ for $C_k = \{d_{k1},\dots,d_{k\,t(k)}\}$.

Denote that the conditions expressed in theorem 8.1.1. (3) and (4) are also necessary when the languages L(RS), L(DRS') are restricted /AABM 80/. Theorem 8.1.1 shows for horizontal decomposition the schema equivalence can be considered as a partition of relations in RS-databases.

Proof. Let d and $d_1,\dots, d_m$ be the following:

$$d = P_1(x_1,\dots,x_n) \vee \dots \vee P_m(x_1,\dots,x_m) ,$$

$$d_i = P(x_1,\dots,x_n) \wedge (d'_{i1} \wedge \dots \wedge d'_{i\,t(i)}) ,$$

where e' is obtained from e by replacement of P_i by P.

(1) We have to prove that for every (DRS',C')-database $M' = (r_1,\dots,r_m)$ there exists a (RS,C)-database $M = (r)$ such that $r = d(M')$. From the hypothesis $C_i \models C$ we conclude that M is a (RS,C)-database. We get also that $r_i \subseteq d_i(M)$.

(2) Given a (RS,C)-database $M = (r)$. Let $M' = (r_1,\dots,r_m)$ where $r_i = d_i(M)$, $1 \leq i \leq m$. Obviously, M' is a (DRS',C')-database. From hypothesis we get $r = d(M')$.

(3) Follows from (1) and (2) using the implication $(RS,C) \prec (DRS',C') \Longrightarrow (RS,C) \leq (DRS',C')$.

(4) We shall prove at first $(RS,C) \prec (DRS',C')$.

Given the (DRS',C') database $M' = (r_1,\dots,r_m)$. Let $r = d(M')$ and $M = (r)$. Obviously M is a (RS,C)-database by hypothesis. Otherwise we get $r_j = d_j(M)$, $1 \leq j \leq m$, and from hypothesis $\models -(C_i) \vee -(C_j)$ for $i \neq j$. Using (2) we obtain that (RS,C) and (DRS',C') are equivalent.

Now we consider some special horizontal decompositions. FD's are the favorite constraints used to decompose schemata.
If a FD $d = X \rightarrow Y$ does not hold in r, then d can not be used to decompose r. However, if the "exceptions" to the FD d are separated from the remaining part of the relation that the main part satisfies d, and hence can be decomposed vertically, according to d. The division of a relation into a subrelation in which d holds and a subrelation in which d does not hold is called /DBPA 83/ the horizontal decomposition according to the goal $\langle X,Y \rangle$, and is formalized below.

A goal is an ordered pair of sets of attributes, $\langle X,Y \rangle$.
We are given two schemes (RS,C), (DRS',C'), $RS = (U,\underline{D},dom)$, $U = \{A_1,\ldots,A_n\}$, $DRS' = RS_1,RS_2$ where $RS_i = (U,\underline{D},dom)$ for $i \in \{1,2\}$.
For

$$d_1 = .(P(x,y,z) \wedge \forall y' \forall z' (P(x,y',z') \rightarrow y = y')),$$

$$d_2 = .(P(x,y,z) \wedge \forall y' \forall z' (P(x,y',z') \rightarrow y \neq y')),$$

$$d = P_1(x,y,z) \vee P_2(x,y,z),$$

the lossless schema transformation $((d_1, d_2),(d))$ describes the horizontal decomposition of (RS,C), according to the goal $\langle X,Y \rangle$.

The horizontal decomposition can be described also in terms of definitions from 4.2.

Let be r_1 the largest X-complete subset of r in which the FD $X \rightarrow Y$ holds and $r_2 = r - r_1$. Then (r) is decomposed into (r_1, r_2).
Formally,

$$r_1 = \{t \in r \mid \forall t' \in r\ (t(X) = t'(X) \rightarrow t(Y) = t'(Y))\} \quad \text{and}$$

$$r_2 = \{t \in r \mid \exists t' \in r\ (t(X) = t'(X) \wedge t(Y) \neq t'(Y))\}.$$

In /DBPA 82/ is shown that the horizontal decomposition, according to a goal, preserves FD's. There, also a new normal form is defined.

A scheme $DS = (RS_1 \ldots RS_m,C)$ with $RS_i = (U = \{A_1,\ldots,A_n\}, \underline{D},dom)$ for i, $1 \leq i \leq m$, is said to be in Goal Normal Form iff for all $X,Y \subseteq \{A_1,\ldots,A_n\}$ and i, $1 \leq i \leq m$, holds $RS_i\colon X \rightarrow Y$ or $RS_i\colon X \not\rightarrow Y$.

Unfortunately, Goal Normal Form can not be used to decompose schemes. Using the goals $\langle X,Y \rangle$ and $\langle Y,X \rangle$ alternatively for horizontal decomposition of a schema $(RS,\emptyset)$ an infinite sequence $((U,\underline{D},dom),\emptyset)$, (RS_1RS_2,C_1), $(RS_1RS_{21}RS_{22},C_2)$, $(RS_1RS_{21}RS_{221}RS_{222},C_3)$ with $RS_{xyz\ldots} = (U,\underline{D},dom)$ can be constructed with no elements being in goal normal form. Therefore, stronger horizontal decompositions are required, one of those is described in detail, below.

8.2. CONDITIONAL FUNCTIONAL DEPENDENCIES

When decomposing a relation horizontally, it may become obvious that some additional constraints must hold in one of the subrelations. For instance, if (in a company) employees can work in some rooms it is obvious that employees who have only one working place will not get more than one telephone number. In /BRPA 83/, a new constraint is introduced for expressing such connections.

Remember, that for a scheme RS = (U = $\{A_1,\ldots A_n\}$,$\underline{U}$,dom) a set X $\subseteq$ U and a RS-database M = (r), a subrelation r' of r is called X-complete iff the tuples not belonging to r' have other X-projections than those belonging to r'.

For X,Y,Z $\subseteq$ U the constraint X --> Y)- X --> Z is called conditional functional dependency (CFD). It means that in every X-complete set of tuples in r in which the FD X --> Y holds, the FD X --> Z must hold, too.

Therefore, a conditional functional dependency can be represented as a second-order formula

$$\forall r' \subseteq r ((\forall t \in r' \forall t' \in r-r' (t(X) \neq t'(X)) \wedge ((r' \models X \dashrightarrow Y) \Longrightarrow (r' \models X \dashrightarrow Z)))) .$$

In our previous example we get the CFD

employee --> room)- employee --> phone .

Assuming that most employees have only one room the part of the relation that include these employees, is almost the entire relation. Now, the horizontal decomposition separates the employees schema and database.

Let RS = (U,$\underline{U}$,dom) be a schema with a set C of FD's. Let X,Y be subsets of U. For every RS-relation r , the restriction $C_{X\to Y}(r)$ for X --> Y of r is the largest X-complete subset of r in which X --> Y holds.

The horizontal decomposition of an RS-database (r) , according to the CFD X --> Y)- X --> Z is a new database (r_1,r_2) with $r_1 = C_{X\to Y}(r)$ and $r_2 = r - r_1$. The decomposition is called nontrivial if $r_1 \neq \emptyset$ and $r_2 \neq \emptyset$.

The horizontal decomposition of a scheme (RS,C) , according to the CFD X --> Y)- X --> Z is the schema DRS' = (RS_1RS_2,C') where RS_1=RS_2 = (U,$\underline{D}$,dom), for every RS-database r there exists one and only one DRS'-database (r_1,r_2) such that $r_1 = C_{X\to Y}(r)$ and $r_2 = r - r_1$,

$$C' = \{RS_i: X' \dashrightarrow Y' \mid X' \dashrightarrow Y' \in C , 1 \le i \le 2\} \cup \{RS_1: X \dashrightarrow Y, RS_2: X \dashrightarrow Z\} \cup \{RS_2: X \not\dashrightarrow Y\} .$$

The afunctional dependency RS_2: X //> Y means that in every non-empty X-complete set of tuples from r_2 on RS_2 from DRS' the FD X --> Y does not hold.

Now we introduce the formal system Γ_{CFD} for axiomatization of the class of conditional functional dependencies.

Formal system Γ_{CFD} .

Axiom XZ --> YZ)- XZ --> Z

Rules

```
     XY --> Z
-----------------
X --> Y )- X -->Z

X --> Y )- X --> Z  ,  X --> Y )- X --> T
-----------------------------------------
            X --> Y )- X --> ZT

X --> Y )- X --> Z  , Z --> T
-----------------------------
    X --> Y )- X --> T

X --> Y )- X --> Z  , X --> Z )- X --> T
----------------------------------------
    X --> Y )- X --> T

X --> Y )- X --> Z  ,  W --> Y )- W --> X  ,  X --> W
-----------------------------------------------------
                 X --> Y )- W --> Z
```

As FD's X --> Y are special CFD's Z --> Z)- X --> Y the use of FD's in these rules is allowed.

<u>Corollary 8.2.1</u>. /BRPA 83/ The formal system Γ_{CFD} is sound for the implication of conditional functional dependencies.

Proof. We only prove the last rule because the others are obviously sound. Let r' be an arbitrary W-complete set of tuples. Since X --> W holds, r' is also X-complete. If W --> Y holds in r' then so does X --> Y by transitivity on X --> W and W --> Y. X --> Y in r' induces X --> Z in r' and W --> Z holds in r' by transitivity.

For the formal system Γ_{CFD} the completeness can be proven introducing the following set S_C(X --> Y) for a FD X --> Y and a set CFD C as the smallest set of FD's with the following properties for a scheme RS = (U,<u>U</u>, dom):

1. X --> Y $\in$ S_C(X --> Y) ;

2. If $T \rightarrow V \in S_C(X \rightarrow Y)$ and $T \rightarrow V$)- $T \rightarrow W \in C$ then
$T \rightarrow W \in S_C(X \rightarrow Y)$;
3. If $X' \rightarrow Y'$, $Y' \rightarrow Z' \in S_C(X \rightarrow Y)$ then
$X'VW \rightarrow WZ' \in S_C(X \rightarrow Y)$ for $V, W \subseteq U$.

Using a property of Armstrong relations the following connection between Γ_{CFD} and $S_C(X \rightarrow Y)$ in /DBPA 83/ it is proven :

(i) If $T \rightarrow V \in S_C(X \rightarrow Y)$ then $C \vdash_{\Gamma_{CFD}} T \rightarrow V$ or $\vdash_{\Gamma_{CFD}} T \rightarrow X$;

(ii) If $T \rightarrow V \in S_C(X \rightarrow Y)$ then
$C \cup \{X \rightarrow Y$)- $X \rightarrow T\} \vdash_{\Gamma_{CFD}} X \rightarrow Y$)- $X \rightarrow V$.

Using these properties we get directly

__Lemma 1__. $C \models X \rightarrow Y$)- $X \rightarrow Z$ iff $X \rightarrow Z \in S_C(X \rightarrow Y)$.

Using this lemma we get a membership algorithm which does not require more than $O(|C|^3 n^2)$ of time and we get

__Theorem 8.2.2__. The formal system Γ_{CFD} is sound and complete for implication of conditional functional dependencies.

A large number of generalizations of conditional functional dependencies is introduced and considered in /DBPA 85/, /DBPA 86/ and other papers of P. De Bra and J. Paredaens.

We are given a scheme $RS = (U = \{A_1,\ldots,A_n\}, \underline{U}, dom)$ and a database $M = (r)$ from $(RS, \emptyset)$.

A set of tuples r' of r is called X-unique if all the tuples of r' have the same X-projection.

The __imposed functional dependency__ $X \rightarrow Y$)- $V \rightarrow Z$ means that the FD $V \rightarrow X$ holds in M, and in every X-complete set of tuples in which the FD $X \rightarrow Y$ holds, the FD $V \rightarrow Z$ must hold, too.

Conditional functional dependencies are special imposed functional dependencies with $V = X$. A goal can be expressed as a trivial CFD $T \rightarrow V$)- $T \rightarrow T$.

The functional dependency implication $X \dashrightarrow Y$)-^{Z} $T \dashrightarrow V$, means that in every Z-complete set of tuples of M in which the FD X --> Y holds, the FD T --> V must hold, too. For Z = X, a functional dependency implication is an imposed functional dependency.

For sets of FD's C_1, C_2, the functional dependency set implication C_1)-^{Z} C_2 means that in every Z-complete set of tuples in M in which all the FD's of C_1 hold, all the FD's of C_2 must hold, too.
The functional dependency implications are special functional dependency set implications in which $_1$ and $_2$ each include only one FD.

The unrestricted functional dependency X --> Y)--^{Z} T --> V holds in H if every Z-complete, Z-unique set of tuples in r in which the FD X --> Y holds, the FD T --> V must hold, too. This dependency is equivalent to the functional dependency implication XZ --> Y)-^{Z} TZ --> V .

A conditional afunctional dependency X --> Y)- X -//-> Z can be defined as the constraint that in an X-complete set the property X --> Y imply the property X --> Z . However this constraint is equivalent to the afunctional dependency X -//-> YZ .

There are also known generalized functional set implications, anti-functional dependencies and anti-functional dependency sets.

For the other dependency classes besides FD's the horizontal decomposition approach can be also useful.

For X,Y,Z $\subseteq$ U the constraint X ->-> Y)- X ->-> Z is called conditional multivalued dependency. It means that in every X-complete set of tuples in which the multivalued dependency X->->Y holds, the multivalued dependency X ->-> Z must hold, too.

For a database scheme DRS = (RS_1 RS_2,C) with RS_1 = (U_1,$_D$,dom1), RS_2 = (U_2,$\underline{D}$,dom2) , U_1 = { A_1,..., A_p }, U_2 = {B_1,..., B_t } , X $\subseteq$ U_1 , Y,Z $\subseteq$ U_2 a conditional inclusion dependency $P_1(X) \subseteq P_2(Y)$)- $P_1(X) \subseteq P_2(Z)$ can be introduced analogously.

These generalizations can be conceived as special representations of logical functions /VASH 78/.

8.3. UNION CONSTRAINTS

Using the results of chapter 6.2 it is possible to axiomatize another class of constraints of horizontal decomposition. The purpose of this chapter is to introduce the notion of union constraints which is a type of database constraints not previously discussed in literature and to show that there exists a sound and complete formal system. In database literature, there is a number of results, both positive and negative, for the existence of finite formal theories. The class of union constraints is the first class of constraints which is known to be axiomatizable and which are not dependencies. By an union constraint it is stated that there exists a cover of the relation with possibilities of "forgetting" some attributes.

We are given a relation scheme RS = (U , $\underline{D}$, dom) where $U = \{A_1,...,A_n\}$ and $X,Y \subseteq U$, $XY = U$. The pair [X,Y] is called <u>union constraint</u>.

A RS-database M = (r) satisfies this constraint if there are subsets r_1, r_2 of r such that $r_1 \cup r_2 = r$ and $r = r_1[X] + r_2[Y]$ (denoted by M ||== [X,Y]).

Only the (full) union constraints [X,Y] with XY = U are of interest because of from r ||== [X,Y] follows Ex(RS',RS)(r[XY]) = r for the subscheme RS' for which is defined r[XY] . Since the validity of a union constraint depends also from $\underline{D}$, only the trivial union constraint [U,U] is a dependency.

Obviously, the constraint [XZ,YZ] can be described with the following formula from L(RS) for disjoint sets X,Y,Z:

$\forall x \forall y \forall z \forall x' \forall y' (P(x,y,z) \rightarrow P(x,y',z) \vee P(x',y,z))$.

<u>Example</u>. Let U = {1,2,3,4}, dom(A) = {0,1} for $A \in U$ and r be the following relation. Then r can be represented by the relations $r_1[\{1,2\}]$ and $r_2[\{1,3,4\}]$

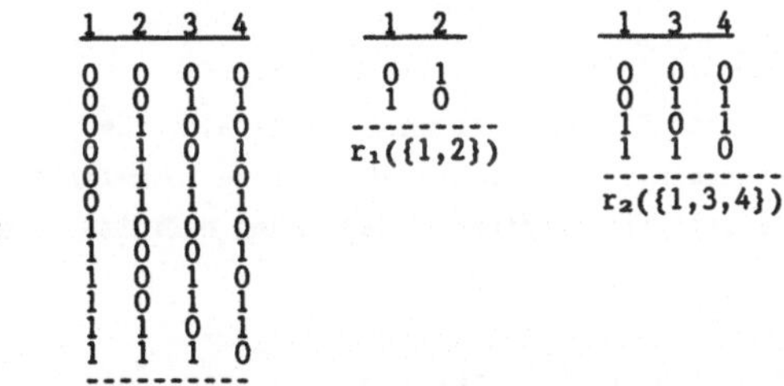

r

1	2	3	4
0	0	0	0
0	0	1	1
0	1	0	0
0	1	0	1
0	1	1	0
0	1	1	1
1	0	0	0
1	0	0	1
1	0	1	0
1	0	1	1
1	1	0	1
1	1	1	0

1	2
0	1
1	0

$r_1(\{1,2\})$

1	3	4
0	0	0
0	1	1
1	0	1
1	1	0

$r_2(\{1,3,4\})$

Let $UCON_2$ be the set of all union constraints of the scheme RS.

Now we can extend the implication also to $UCON_2$.

Let C be a set of union constraints and $[X,Y] \in UCON_2$.

From C follows $[X,Y]$ (denoted by $C \models [X,Y]$) if for every RS-database $M = (r)$ from $r \,||\!\!== C$ follows $r \,||\!\!== [X,Y]$.

There exists an equivalence between $UCON_2$ and $JDEP_2$. For $C \subseteq JDEP_2$ and $\Phi \subseteq UCON_2$ we define

$JDEP_2(\Phi) = \{(X,Y) \mid [X,Y] \in \Phi\}$ and

$UCON_2(C) = \{[X,Y] \mid (X,Y) \in C\}$.

Using a new predicate P' which is defined as $P'(u) \dashrightarrow -P(u)$ we get

<u>Corollary 8.3.1</u>. For $C \subseteq JDEP_2$, $(X,Y) \in JDEP_2$

$C \models (X,Y)$ iff $UCON_2(C) \models [X,Y]$.

Now we define the formal system Γ_{UC} .

Formal system Γ_{UC} .

Axiom $[U,U]$.

Rules

(1) $\dfrac{[X,Y]}{[V,W]}$ if $(X,Y) \leq (V,W)$

(2) $\dfrac{[X_1,X_2] \;,\; [Y_1,Y_2]}{[X_1 \cap Y_1, Y_2]}$ if $X_1 \cap X_2 \subseteq Y_1$, $X_2 \subseteq Y_2$.

Using the above corollary and the result of chapter 5.1 we get

<u>Theorem 8.3.2</u>. The system Γ_{UC} is sound and complete for implication of union

constraints.

Since union constraints are not definite formulas and all the other presented and known classes of constraints are classes of definite formulas this result is the first axiomatization result for constraints not being definite formulas.

Example. Let U = {BAR, DRINKER, BEER} and r be a relation on U where only first class bars which serves any sort of beers and also bars which are sometimes frequented by any drinker are represented. Then r can be represented by the relation $r_1[\{BAR, DRINKER\}]$ of first class bars and by the relation $r_2[\{BAR, BEER\}]$ of frequented bars.

9. THE RELATIONSHIP BETWEEN DEPENDENCY CLASSES

In the previous chapters, more than 80 different dependency classes are introduced and considered. In /THAL 86/, more than 600 different references to papers on dependency theory are given. By some authors it was noticed that dependency theory is in a chaotic state. This book should be understood as an attempt to present the most important results on dependency theory. The usefulness of such a great number of different constraints is an open problem. But the variety can be explained as follows:

1. Each new type represents a certain type of semantic constructions.
2. Many types are connected with normalization and decomposition theory of databases.
3. Some types are generalizations of the previous ones.
4. Some types are introduced as special tools for manipulation and control of data.
5. Some types improve the utilization of projections of relations or of partition of relations.

But the large number of different dependency classes also demonstrates the incompleteness of the theory and requires a systematized extension of the presented results. In this book, for examination of different types, only three characteristics were of interest: conditions for existence; semantic restrictions; connections with other types. It is only something known about comparisons of practical applicability of different types. As noticed in /DEAD 85/, in practice these dependency classes are never used to the same extend. Because of their easy nature, functional dependencies are widely employed and form the basis for identifying tuples and data.

This book aims at an attempt to systematize the dependency theory. In /THYA88/, the presented theory is used for proposing a general constraint theory for value-oriented database models based on the Higher-order Entity-Relationship Model. The following figures depict the relation between the different types of dependencies described. An arrow K --> L means that the dependencies of type L can be described in terms of type K . Any dependency of type L logically implies a dependency of type K . There always exists some dependency of type K which is equivalent to a given dependency of type L . Different classes are equal. They are presented together like synonyms.

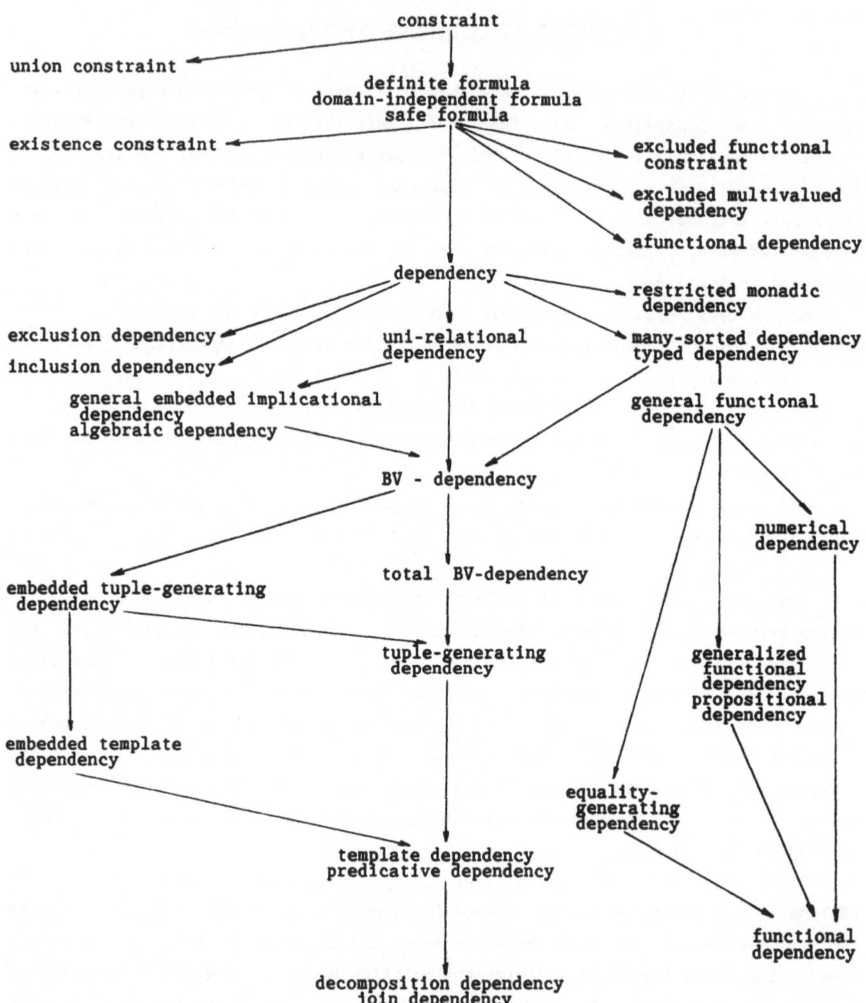

Figure 1. The general picture.

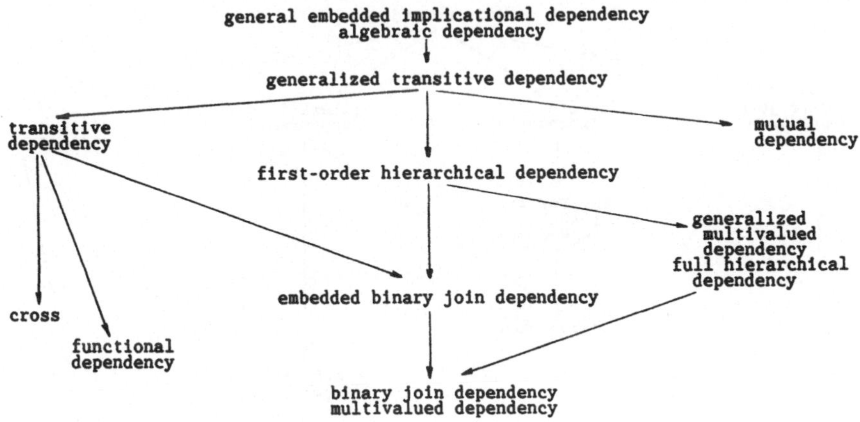

Figure 2. The algebraic dependencies.

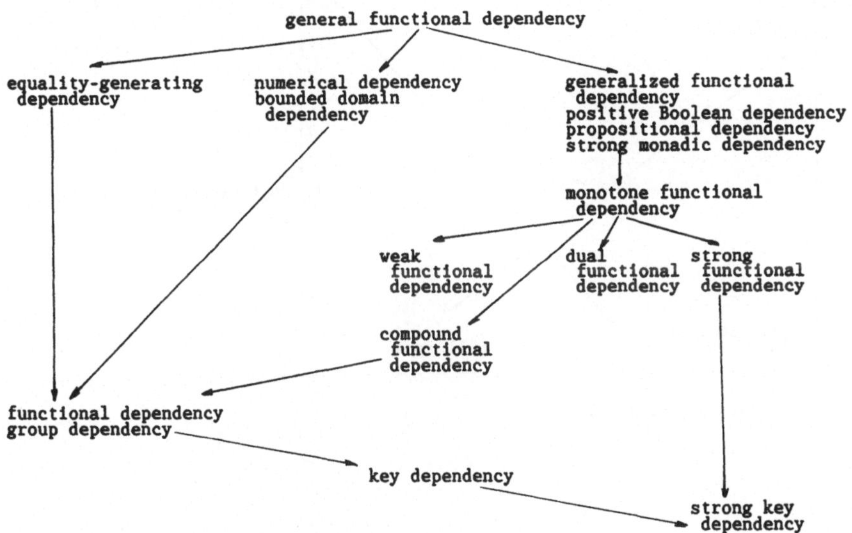

Figure 3. The functional dependencies.

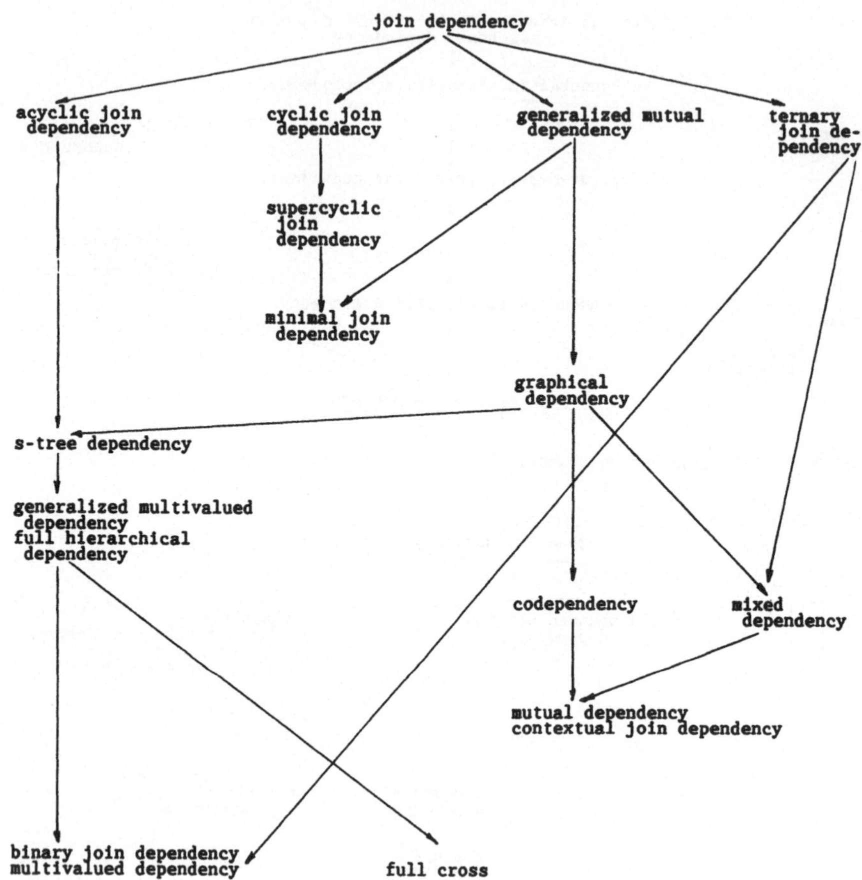

Figure 4. Join dependencies.

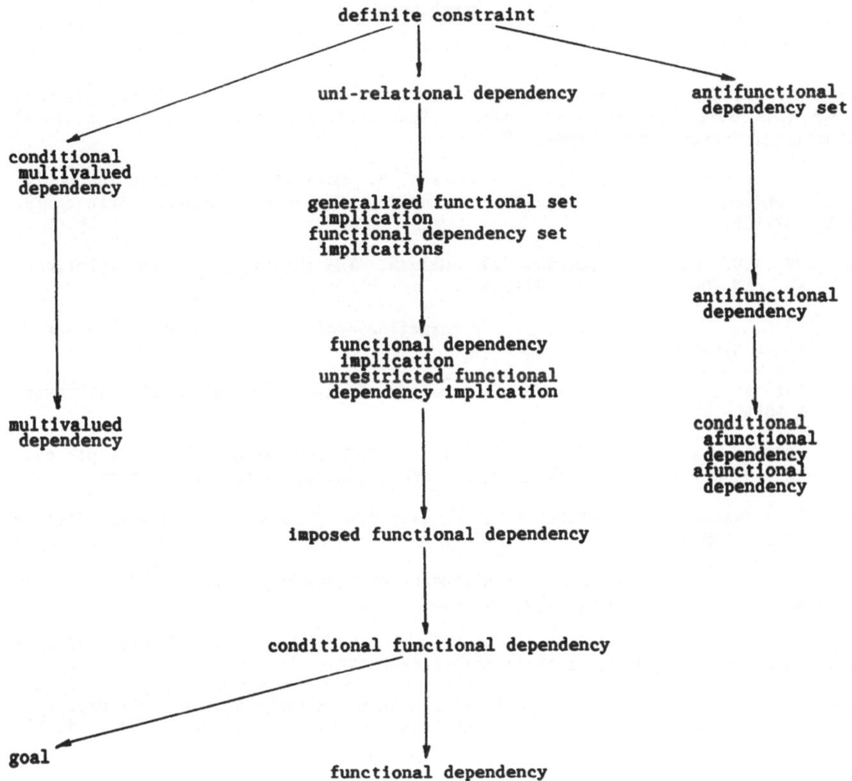

Figure 5. Horizontal decomposition dependencies.

REFERENCES

/AABM 80/ P. Atzeni, G. Ausiello, C. Batini, M. Moscarini, Conceptual relations among relational database schemata. Technical report R-80-32, Instituto di Automatica, University of Rome, 1980.

/AABM 82/ P. Atzeni, G. Ausiello, C. Batini, M. Moscarini, Inclusion and equivalence between relational database schemata. Theoretical Computer Science 19, 1982, 267-285.

/ABU 79/ A.V. Aho, C. Beeri, J.D. Ullman, The theory of Join in relational database. ACM TODS 4,3, 1979, 297-314.

/ABVI 85/ S. Abiteboul, V. Vianu, Transactions and integrity constraints. Proc. of Database Systems, 1985, 193-204.

/AHUL 79/ A. Aho, J.D. Ullman, Universality of data retrieval languages. Proc. 6th ACM POPL, 1979, 110-117.

/ALFT 88/ S. Al Fedaghi, B. Thalheim, Logical foundations for two-tuple constraints in the relational database model. 60 p. Submitted for publication.

/ANSI 75/ ANSI/X3/SPARC, Study group on data base management systems, Interim Report, EDT, ACM SIGMOD records, 7, 2, 1975.

/ARDE 80/ W.W. Armstrong, C. Delobel, Decompositions and functional dependencies in relations. ACM TODS, 5,4, 1980, 404-430.

/ARM 74/ W.W. Armstrong, Dependency structures of data base relationships. Information processing 74, North-Holland, Amsterdam, 1974, 580-583.

/ARMS 66/ D.B. Armstrong, On Finding a Nearly Minimal Set of Fault Detection Tests for Combinatorial Logic Nets. IEEE Trans. on Electr. Comput., 1966, EC-15, 66-73.

/ARSM 81/ S.K. Arora, K.C. Smith, A graphical interpretation of dependency structures in relational data bases. Int. J. Comp. and Inf. Sci., 1981, v. 10, No. 3, 187-213.

/ATMO 84/ P. Atzeni, N.M. Morfuni, Functional dependencies in relations with null values. Information Processing Letters, 18, 14May84, 233-238.

/AUBM 80/ G. Ausiello, C. Batini, M. Moscarini, On the equivalence among database schemata, Proc. Int. Conference on Data Bases, Aberdeen, 1980, Chapter 3, 34-46.

/AUAS 83/ G. Ausiello, A.D. Atri, D. Sacca, Graph algorithms for functional dependency manipulation. J. ACM 30, 1983, 752-766.

/BARI 84/ F. Bancilhon, P. Richard, A sound and complete axiomatization of embedded cross dependencies. Theoretical Computer Science 34, 1984, 343-350.

/BASP 81/ F. Bancilhon, N. Spyratos, Independent components of data bases. 7th

Inf. Conf. on VLDB, 1981, 398-408.

/BDFS 84/ C. Beeri, M. Dowd, R. Fagin, R. Statman, On the structure of Armstrong relations for functional dependencies. Journal of ACM, Vol.31, No.1, January 1984, 30-46.

/BDHF 80/ A. Bekessy, J. Demetrovics, L. Hannak, P. Frankl, G. Katona, On the number of maximal dependencies in a database relation of fixed order. Discrete Math. 1980, 30, 83-88.

/BDKK 88/ G. Burosch, J. Demetrovics, G.O.H. Katona, D.J. Kleitman, A.A. Saposhenko, On the number of databases and closure operations. To appear in J. Comp. Sci.

/BEBE 79/ C. Beeri, P.A. Bernstein, Computational problems related to the design of normal forms in relational schemes. ACM TODS 4, 1, 1979, 30-59.

/BEBL 85/ J. Berman, W.J. Blok, Positive Boolean dependencies. University of Chicago, Research Reports in Computer Science, No.5, June, 1985.

/BEDE 79/ A. Bekessy, J. Demetrovics, Contribution to the theory of data base relations. Discrete Math. 1979, 27, 1-10.

/BEHO 81/ C. Beeri, P. Honeyman, Preserving functional dependencies. SIAM J. Computing 10, 3, 1981, 647-656.

/BEKI 86/ C.. Beeri, M. Kifer, An integrated approach to logical design of relational database schemes. ACM TODS, 11, 1986, 159-185.

/BENE 88/ K. Benecke, On hierarchical normal forms. Proc. MFDBS-87, Dresden 1987, LNCS 305, p. 10-19.

/BEVA 81/ C. Beeri, M.Y. Vardi, On the properties of join dependencies. Advances in Database Theory (eds: H. Gallaire, J. Minker, J.M. Nicolas), New York, Plenum Press, 25-72, 1981.

/BEVA 84/ C. Beeri, M.Y. Vardi, A property for data dependencies. Journal of ACM, 31, 4, 1984, 718-741.

/BEVA 85/ C. Beeri, M.Y. Vardi, Formal systems for join dependencies. Theoretical Computer Science 38, 1985, 99-116.

/BFH 77/ C. Beeri, R. Fagin, J.H. Howard, A complete axiomatization for functional and multivalued dependencies in database relations. Proc. ACM SIGMOD, Toronto, 1977, 47-81.

/BFMY 83/ C. Beeri, R. Fagin, D. Maier, M. Yannakakis, On the desirability of acyclic database schemes. Journal of ACM, 30, 3 1983., 479-513.

/BIBD 79/ J. Biskup, P.A. Bernstein, V. Dayal, Synthesizing independent data base schemes. Proc. ACM SIGMOD Conf., 1979, 143-151.

/BIBR 83/ J. Biskup, H.H. Brüggemann, Designing acyclic database schemes. Advances in Database Theory, Vol. II (eds. H. Gallaire, J. Minker, J.-M. Nicolas), Plenum-Press, 1983, 3-26.

/BISK 78/ J. Biskup, On the complementation rule for multivalued dependencies in data base relations. Acta informatica 10, 1978, 297-305.

/BISK 83/ J. Biskup, A foundation of Codd's relational may-be operations. ACM TODS 8, 1983, 608-636.

/BROS 80/ M.L. Brodie, J.W. Schmidt, Standardization and the relational approach to data bases: an ANSI Task Group Status Report. 6th Int. Conf. VLDB, 1980, 326-328.

/BÖRG 85/ E. Börger, Berechenbarkeit, Komplexität, Logik. Vieweg, Braunschweig 1985.

/BUDK 87/ G. Burosch, J. Demetrovics, G.O.J. Katona, The poset of closures as a model of changing databases. Order 4, 1987, 127-142.

/BUOR 86/ W. Buszkowski, E. Orlowska, On the logic of database dependencies. Bull. Polish Academy of Sciences, Vol. 34, 5-6, 1986, 345-354.

/BVAR 84/ C. Beeri, M.Y. Vardi, Formal systems for tuple and equality generating dependencies. SIAM J. Computing, 13, 1, 1984, 76-98.

/CASA 81/ M. A. Casanova, The theory of functional and subset dependencies over relational expressions. Dep. de Inf. Rep. 3/81, Pont. Univ. Cat.., Rio de Janeiro, Jan. 1981.

/CAVI 83/ M.A. Casanova, V.M.P. Vidal, Towards a sound view integration methodology. 2nd ACM SIGMOD Symposium on Principles of Databse systems, 1983, 36-47.

/CFP 84/ M.A. Casanova, F. Fagin, C.H. Papadimitrou, Inclusion dependencies and their interaction with functional dependencies. JCSS, Vo.28, No.1, February 1984, 29-59.

/CEGT 88/ S. Ceri, G. Gottlob, A. Tanca, Logic Programming and databases. Springer 1988.

/CHEN 76/ P.P. Chen, The Entity-Reltationship Model: Towards a unified views of data. ACM TODS, 1, 1, 76, 9-26.

/CHEN 84/ P.P. Chen, An algebra for a directional binary Entity-Relationship Model. Proc. 1st IEEE Intl. Conf. on data Engineering, Los Angeles 1984, 37-40.

/CHHE 88/ E.P.F. Chan, H.J. Hernandez, Independence reducible database schemes. ACM SIGACT-SIGMOD-SIGART 1988 Conf., 163-173.

/CHKE 73/ C.C. Chang, H.J. Keisler, Model theory. Amsterdam, North-Holland 1973.

/CHLE 73/ C.L. Chang, R.C.T. Lee, Symbolic logic and mechanical theorem proving. Academic press, New York, 1973.

/CHLM 81/ A.K. Chandra, H.R. Lewis, J.A. Makowsky, Embedded implicational dependencies and their inference problem. ACM Symp. on Theory of Computing, 1981, 342-354.

/CHVA 83/ A.K. Chandra, M.Y. Vardi, The implication problem for functional and inclusion dependencies is undecidable. Technical report, Stanford University, Dept. of Comp. Sci., March 1983.

/CODD 70/ E.F. Codd, A relational model for large shared data banks. Comm. ACM 13, 6,1970, p. 197-204.

/CODD 71/ E.F. Codd, Further normalization of the database model, In: Courant Inst. Comp. Sci. Symp. 6, Data Base Systems, Prentice Hall, Englewood Cliffs 1971, p. 33-64.

/CODD 72/ E.F. Codd, Relational completeness of data base sublanguages. In: Data base systems (ed. R. Rustin), Prentice Hall, Englewood Cliffs, NJ, 1972, 65-98.

/CODD 79/ E.F. Codd, Extending the relational database model to capture more meaning. ACM TODS 4, 4, 1979, 397 - 434.

/CODD 81/ E.F. Codd, Data models in database management. Proc. Workshop on Data Abstraction, Databases and Conceptual Modelling, SIGPLAN Notices, Vol. 16, 1, 1981, 112 - 114.

/CODD 82/ E.F. Codd, Relational databases: A practical foundation for productivity. Comm. ACM, 25, 2, Febr. 82, 109-117.

/CODD 86/ E.F. Codd, Missing Information (Applicable and Inapplicable) in Relational Databases. SIGMOD Record, Vol. 15, No. 4, Dec. 1986, 53 -78.

/COKA 83/ S.S. Cosmadakis, P.C. Kanellakis, Functional and inclusion dependencies - A graph theoretic approach. Technical Report Cs-83-21, Brown University, Dept. of Comp. Sci.

/CRAI 67/ A. Craig, Modus ponens and derivation from Horn formulas. Zeitschrift fur Mathematische Logik und Grundlagen der Mathematik 13, 1967, 33-54.

/CZED 81/ G. Czedli, On dependencies in the relational model of data. EIK 17 (1981), 2/3, 103-112.

/DAPA 88/ Dawson K.S., Parker L.M.P., From entity-relationship diagrams to fourth normal form: A pictorial aid to analysis. The Computer Journal, 31, 3, 1988, p. 258-268.

/DBPA 82/ P. De Bra, J. Paredaens, Horizontal decompositions for handling exceptions to functional dependencies. Report 82-20, University of Antwerp, Dept. of Mathematics, 1982.

/DBRA 83/ P. De Bra, J. Paredaens, Conditional dependencies for horizontal decompositions. LNCS 154, 1983, 67-82.

/DBRA 85/ P. De Bra, Horizontal decompositions based on functional dependency-set-implications. Report Universiteit Antwerpen, Dept. of Mathematics, 85-35, Oct. 1985.

/DBRA 86/ P.De Bra, Functional dependency implications, including horizontal decompositions. Submitted report to Mathematical fundamentals of Database Systems, Dresden, 1986.

/DEAD 85/ C. Delobel, M. Adiba, Relational database systems. North-Holland, Amsterdam 1985.

/DECA 85/ C. Delobel, R.G. Casey, Decopositions of a data base and the theory of Boolean switching functions. IBM J. Res. Dev. 17, 1973, 374-386.

/DEFK 85/ J. Demetrovics, Z. Füredi, G.O.H. Katona, Minimum matrix representations of closure operations. Discrete Applied Mathematics 11, 1985, 115-128.

/DEGY 81/ J. Demetrovics, Gy. Gyepesi, On the functional dependency and some generlizations of it. Acta Cybernetica 5 (1981), 295-305.

/DEGY 83/ J. Demetrovics, Gy. Gyepesi, A note on minimal matrix representation of closure operations. Combinatorica 1983, 3, 2, 177-179.

/DEKA 83/ J. Demetrovics, G.O.H. Katona, Combinatorial problems of database models. Colloquia Mathematica Societatis Janos Bolyai 42, Algebra, Cominatorics and Logic in Computer Science, Gyor (Hungary), 1983, 331-352.

/DELM 88/ J. Demetrovics, L.O. Libkin, I.B. Muchnik, Functional dependencies and the semilattice of closed classes. Presented to MFDBS 89, appears in LNCS 364.

/DELO 73/ C. Delobel, Contributions theoretiques a la conception d'un systeme d'information. These d'Etat, Universite de Grenoble, 1973.

/DELO 78/ C. Delobel, Normalization and hierarchical dependencies in the relational data model. ACM TODS 1978, 3, 3, 201-222.

/DELO 80/ C. Delobel, An overview of the relational data theory. IFIP-1980, 413-426.

/DEME 78/ J. Demetrovics, On the number of candidate keys. Information Processing Letters, 1978, 7, 6, 266-269.

/DEME 79/ J. Demetrovics, On the Equivalence of Candidate Keys with Sperner Sets. Acta Cybernetica, Vol. 4, No. 3, Szeged, 247 -252.

/DEME 80/ J. Demetrovics, Candidate keys and antichains. SIAM J. on Algebraic and Discrete Methods, 1980, 1, 92.

/DEME'80/ J. Demetrovics, Relacios adatmodell logikai es structuralis vizsgalata. Tanulmanyok 114, 1980, 1-94.

/DETH 87/ J. Demetrovics, V.D. Thi, Relations and minimal keys. Acta Cybernetica, 1988, 8, 3, 279-285.

/DETH 88/ J. Demetrovics, V.D. Thi, Some results about functional dependencies. Acta Cybernetica, 8, 3, 1988, 273-278.

/DIPA 69/ R. Di Paola, The recursive unsolvability of the decision problem for the class of definite formulas. Journal of ACM 16, 2, 1969, 324-327.

/DRGO 79/ B. Dreben, W.B. Goldfarb, The decision problem - solvable classes of quantificational formulas. Addison-Wesley, New York 1979.

/DYBJ 84/ P. Dybjer, Some results on the deductive structure of join depedencies. Theoretical Computer Science 33, Sept. 84, 95-105.

/FAG 77/ R. Fagin, Multivalued Dependencies and a new normal form for relational databases. ACM Tods 2, 3, 1977, 262-278.

/FAG 80/ R. Fagin, Horn clauses and database dependencies. Proc. 12th Ann. Symp. on the theory of computing, 1980, 123-134.

/FAG 81/ R. Fagin, A normal form for relational data bases that is based on domains and keys. ACM TODS , 1981, 6, 3, 387-415.

/FAG 82/ R. Fagin, Armstrong Databases, Research report IBM Res. Lab., RJ 3440(40926) 4/5/82, San Jose 1982.

/FAG 83/ R. Fagin, Degrees of acyclicity for hypergraphs and relational database schemes. IBM Res. Report RJ 3330 (39949), 11/25/81, 1983.

/FERN 84/ M.C. Fernandez, Determining the normalization level of a relation on the basis of Armstrong's axioms. Computers and Artificial Intelligence, 3, 1984, 495-504.

/FIGU 84/ P.C. Fischer, D. van Gucht, Weak multivalued dependencies. ACM SIGACT/SIGMOD principles of database systems, April 1984, 266-274.

/FMUY 83/ R. Fagin, D. Maier, J.D. Ullman, M. Yannakakis, Tools for template dependencies. SIAM J. Comput., 12, 1, 1983, 30-59.

/FROS 85/ R.A. Frost, Formalizing the notion of semantic integrity in database and knowledge systems. Proc. 5th British Nat. Conf. on Databases, 105-127.

/FSTG 85/ P.C. Fischer, L.V. Saxton, S.J. Thomas, D. Van Gucht, Interactions between depedencies and nested relational structures. J. Computer and System Sciences 31, 1985, 343-354.

/GAJO 79/ M.R. Garey, D.S. Johnson, Computers and Intractability: a Guide to the theory of NP-completeness. Freeman, 1979.

/GAMN 84/ H. Gallaire, J. Minker, J.M. Nicolas, Logic and databases: a deductive approach. Computing Surveys 16, June 1984, 153-185.

/GAYE 88/ S.K. Gadia, C.-S. Yeung, A generalized model for a relational temporal database. Proc. ACM SIGMOD 1988, June 1988, Chicago, p. 251-259.

/GERO 81/ J. Getta, S. Romanski, Group depedencies in relational data bases. Arch. Automat. Telemech. 26, 1981, 3, 365 -372.

/GIZA 82/ S. Ginsburg, S.M. Zaiddan, Properties of functional dependency families. Journal ACM, 1982, 678-698.

/GOLD 81/ B.S. Goldstein, Formal properties of constraints on null values in relational databases. Technical report 80-013 SUNY at Stony Brook, Dept. of Computer Science, 1981.

/GOSS 88/ G. Gottlob, M. Schefl, M. Stumptner, On the interaction between transitive closure and functional dependencies. Submitted to MFDBS-89, Wien 1988.

/GOTA 84/ N. Goodman, Y.G. Tay, A characterization of multivalued dependencies equivalent to a join dependency. Information Processing Letters 18, 1984, 261-266.

/GOTT 87/ G. Gottlob, On the size of nonredundant FD-covers. Information Processing Letters, 24, 6, 6 Apr. 1987, 355-360.

/GOTT'87/ G. Gottlob, Computing covers for embedded functional dependencies. ACM SIGACT-SIGMOD-SIGART Symp. 1987, 58-69.

/GRAN 79/ J. Grant, Null values in a relational data base. Information processing letters, 6,5, 1979, 156 -157.

/GRMV 86/ M.H. Graham, A.O. Mendelzon, M.Y. Vardi, Notions of dependency satisfaction. J. ACM 33, 1, 1986, 105-129.

/GPT 80/ O.Ju. Gorstchinskaja, S.W. Petrow, L.A. Tenembaum, Rasloshenije otnoschenij i logitschekaja projektirowka bas dannyx. Awtomatika i telemechanika 1980, 2, 159-166; 3, 152-160. (In Russian).

/GRJA 82/ J. Grant, B.E. Jacobs, On the family of generalized dependency constraints. Journal of ACM 29,4, 1982, 986-997.

/GRMI 85/ J. Grant, J. Minker, Inferences for numerical dependencies. Theoretical Computer Science 41, 1985, 271-287.

/GULE 82/ Y. Gurevich, H.R. Lewis, The inference problem for template dependencies. Proc. 1st Symp. PODS, 1982, 199-204.

/GURE 76/ Y. Gurevich, The decision problem for standard classes. Journal of Symbolic Logic 41(1976), 460-464.

/GURE 84/ Y. Gurevich, Towards logic tailored for computational complexity. LNM 1104, Springer-Verlag, Berlin 1984, 175-216.

/GYPA 83/ M. Gyssens, J. Paredaens, Another view of functional and multivalued dependencies in the relational database model. Int. J. Computer and Information Sciences 12, Aug 1983, 247-267.

/GYPA 86/ M. Gyssens, J. Paredaens, On the decomposition of join dependencies. Advances in Computing Research 3, 1986, 69-106.

/GYSS 86/ M. Gyssens, On the complexity of join dependencies. ACM TODS 1986, 11, 1, 81-108.

/HAFA 86/ Y. Hanatani, R. Fagin, A simple characterization of database dependency implication. Information Processing Letters, 22, 30 May 1986, 281-283.

/HEGN 88/ S.J. Hegner, Decomposition of relational schemata into components defined by both projection and restriction. ACM SIGACT-SIGMOS-SIGART Sym. 1988, 174-183.

/HONE 82/ P. Honeyman, Testing satisfaction of functional dependencies. Journal ACM 1982, 668-677.

/HOTH 86/ Ho Thuan, Contribution to the theory of relational databases. Manuscript, Budapest 1986.

/HTLB 84/ Ho Thuan, Le Van Bao, Some results about keys of relational schemes. Acta Cybernetica, Tom 7, Fasc. 1, Szeged, 1984, 99-113.

/HUGI 83/ R. Hull, S. Ginsburg, Order Dependencies in the relational model. Theoretical Computer Science 26, 1983, 149-195.

/HULL 84/ R. Hull, Finitely specifiable implicational dependency families. J. ACM 31, 1984, 210-226.

/IMLI 82/ T. Imielinski, W. Lipski Jr., A systematic approach to relational database theory. ICS PAS Reports 457, Warszawa, 1982.

/IMLI 83/ T. Imielinski, W. Lipski, Incomplete information and depedencies in relational databases. SIGMOD REC., 1983, 13, 4, 178-184.

/JACO 82/ B. Jacobs , On database logic. J. ACM, 29, 2, 1982, p. 310-332.

/JAJO 86/ S. Jajodia, Recognizing multivalued dependencies in relation schemes. Computer Journal, 29, Oct. 1986, 458-459.

/JALU 80/ S.W. Jablonski, O.B. Lupanow, Diskrete Mathematik und mathematische Fragen der Kybernetik, Akademie-Verlag Berlin, 1980.

/JAES 82/ G. Jaeschke, H.J. Schek, Remarks on the algebra of nonfirst-normal-form relations. Proc. First ACM SIGACT-Sigmod Symposium on Principles of Database systems, 1982, 124-138.

/JANT 88/ K.-P. Jantke, Inductive Inference of Functional Dependencies. Report Humboldt University Berlin, ORZ, Aug. 1987.

/JARO 83/ A. Jankowski, C. Rauscer, Logical foundations approach to users domain restriction in databases. Theoretical Computer Science 23, March 1983, 11-26.

/JAPA 79/ D. Janssens, J. Paredaens, General depedencies. Universitaire instelling Antwerpen, Dept. Wiskund, Report 79-35.

/JGK 70/ S.W. Jablonski, G.P. Gawrilow, W.B. Kudrjavcev, Boolesche Funktionen und Postsche Klassen, Akademie-Verlag, Berlin 1970.

/KANE 80/ P.C. Kanellakis, On the computational complexity of cardinality constraints in relational databases. Information processing letters 11, 2, 1980, 98-101.

/KATS 84/ H. Katsuno, When do non-conflict free multivalued dependency sets appear. Information Processing Letters 18, Feb. 84, 87-92.

/KATY 79/ Y. Kambayashi, K. Tanaka, S. Yajima, Semantic aspects of data dependencies and their application to relational database design. Proc. COMPSAC, Nov. 1979, 398-403.

/KAYT 80/ Y. Kambayashi, S. Yajima, K. Tanaka, Problems of relational database design. LNCS 132, Data base design techniques I, p. 172-218.

/KCV 83/ P.C. Kanellakis, S.S. Cosmadakis, M.Y. Vardi, Unary Inclusion dependencies have polynomial time inferance problems. Technical report CS-83-09, Brown University, Dept. of Comp.Sci.

/KELL 85/ A.M. Keller, Set-theoretic problems of null completion in relational databases. Information Processing Letters 22, 28 April 1986, 261-265.

/KLIP 83/ B. Klipps, Ein allgemeiner Abhangigkeitsbagriff fur relationen und seine Axiomatisierung. Preprint WPU Rostock, Mathematik, Juni 1983.

/KOBA 85/ I. Kobayashi, An overview of database management technology. In: Advances in Information System Science" (ed. J.T. Tou), Vol.9, Plenum Press, New York, 1985.

/KOBA 86/ I. Kobayashi, Databases and conceptual schemata: A formal framework, Proc. Conf. VLDB, 1986, Kyoto, 3-23.

/KOBA'86/ I. Kobayashi, Losslessnee and semantic correctness of database schema transformation: Another look of schema equivalence. Inform. Systems, 11, 1, 1986, p. 41-59.

/KOBA"86/ I. Kobayashi, Classification and transformation of binary relationship relation schemata. Inform. Systems, 11, 2, 1986, p. 109-122.

/KOSI 86/ H.F. Korth, A. Silberschatz, Database System Concepts. Mc Graw-Hill Book Company, New York 1986.

/KOST 82/ A.V. Kostochka, On the maximum size of a filter in the n-cube. Prepared for publication, 1982.

/KOST 84/ A.W. Kostotschka, O maksimalnoj moschnosti graniza filtra v n-mernom kube. Diskretnij Analiz, 41, 49-61, Novosibirsk 1984 (in Russian).

/KRKR 67/ G. Kreisel, J.L. Krivine, Elements of mathematical logic; theory of models. Amsterdam, North-Holland, 1967.

/KSCH 25/ K. Knopp, I. Schur, Elementare Beweise einiger asymptotischer Formeln der additiven Zahlentheorie. Mathematische Zeitschrift 24 (1925), 559-574.

/LAMG 83/ K. Laver, A.O. Mendelzon, M.H. Graham, Functional dependencies on cyclic database schemes. Proc. ACM SIGMOD, May 1983, San Jose, 79-91.

/LERV 88/ C. Lecluse, P. Richard, F. Velez, O_2, an object-oriented data model. Proc. ACM SIGMOD, Chicago, June 1988, p. 424-433.

/LIEN 79/ Y.E. Lien, Multivalued dependencies with null values in relational databases. Proc. 5th VLDB, Rio de Janeiro, 1979, 61-66.

/LIEN 82/ Y.E. Lien, On the equivalence of database models. J. ACM 29, 2, April 1982, 333-363.

/LIPS 81/W. Lipski Jr., On database with incomplete information, Journal of ACM, 28, 1, 1981, 41-70.

/LUOS 78/ C.L. Lucchesi, S.L. Osborn, Candidate Keys for Relations. JCSS 17, 1978, 270 - 279.

/MAI 83/ D. Maier, The theory of relational databases. Computer Science Press, Rockville, MD, 1983.

/MAKO 81/ J.A. Makovsky, Characterizing data base dependencies. Proc. ICALP 81, LNCS 1981, 115, 86-97.

/MAMR 85/ J. Makowsky, V.M. Markowitz, N. Rotics, Entity-relationship consistency for relational schemes. Technical report 392, Technion, Haifa, 1985.

/MAPI 82/ F. Manola, A. Pirotte, CQLF - a query language for CODASYL-type databases. Proc. ACM SIGMOD Intl. Conf. on Management of Data, Florida 1982, p. 94-103.

/MARA 82/ H. Mannila, K.-J. Räihä, On the relationship between minimum and optimum covers for a set of functional dependencies. Res. Rep. C-1982-51, University of Helsinki, 1982.

/MARA 86/ H. Mannila, K.-J. Räihä, Inclusion dependencies in database design. Proc. Int. Conf. Data Engineering, 1986, 711-718.

/MAVA 85/ J.A. Makowsky, M.Y. Vardi, On the expressive power of data dependencies. Research report Swiss Federal Institute of Technology, 1985.

/MEMA 79/ A.O. Mendelzon, D. Maier, Genralized mutual dependencies and the decomposition of database relations. Proc. 1979 VLDB, 75-82.

/MEND 79/ A.O. Mendelzon, On axiomatizing multivalued dependencies in relational databases. J. ACM 1979, 26, 1, 37-44.

/MINI 83/ J. Minker, J.M. Nicolas, On recursive axioms in deductive data bases. Information Systems 8, 1, 1983, 1-13.

/MITC 83/ J.C. Mitchell, The implication problem for functional and inclusion dependencies. Information and Control, Vol.53, No.3, March 1983, 145-173.

/MMS 79/ D. Maier, A.O. Mendelson, Y. Sagiv, Testing implications of data dependencies. ACM TODS 4, 4, 1979, 455-469.

/MSTA 66/ A.A. Mitalauskas, V.A. Statusljavistschus, Lokalnije predelnije teoremi i asymptotitscheskije rasloshenija dlja summ nesawisimich reschettschatich slutschanjich welitschin. Litowskij matematitischeskij sbornik, 1966, t. 6, No.4, 569-583.

/MSY 81/ D. Maier, Y. Sagiv, M. Yannakakis, On the complexity of testing implications of functional and join dependencies. Journal of ACM, 28, 4, 1981, 680-695.

/MWIS 77/ F.J. Mac Williams, N.J.A. Sloane, The theory of error-correcting codes. North-Holland, Amsterdam 1977.

/NCHT 87/ N. Cat Ho, B. Thalheim, On Semantic and Syntactic Issues of Null Values in the Relational Model of Data Bases. Submitted for publication 1987.

/NICO 78/ J.-M. Nicolas, First-order logic formalization for functional, multivalued and mutual dependencies. Proc. 1978, ACM SIGMOD, 40-46.

/NIDE 83/ J.-M. Nicolas, R. Demolombe, On the stability of relational queries, In: Logical Bases for databases, Toulouse, 1982.

/PAGU 88/ J. Paredaens, D. Van Gucht, Possibilities and limitations of using flat operators in nested algebra expressions. Proc. ACM SIGACT-SIGMOD-SIGART Symp. PODS, March 1988, Austin, p. 29-38.

/PAPA 86/ C.Papadimitriou C., The theory of database concurrency control. Computer Science Press, Rockville (MD), 1986.

/PAPA 80/ D.S. Parker, K. Parsaye-Ghomi, Inferences involving embedded multivalued dependencies and transitive dependencies, Proc. ACM SIGMOD, 1980.

/PAR 80/ J. Paredaens, The iteraction of integrity constraints in an information system. Journal of Computer and System Sciences, 20, 3, 1980, 310-327.

/PARE 80/ J. Paredaens, Transitive dependencies in a database scheme. RAIRO Inform., 1980, 14, 1, 149-165.

/PARE 82/ J. Paredaens, A universal formalism to express decompositions, functional dependencies and other constraints in a relational data base. Theor. Comp. Sci., 1982, 19, 2, 143-163.

/PAWL 73/ Z. Pawlak, Mathematical foundations of information retrieval. CC PAS Reports 101, Warszawa, 1973.

/PDGG 88/ J. Paredaens, De Bra P., Gyssens M., Van Gucht D., Structures in the relational database model. Springer, Heidelberg 1988.

/PETR 89/ S.V. Petrov, Finite axiomatization of languages for representation of system properties: Axiomatization of dependencies. Information Sciences 47, 1989, 339-372.

/REI 84/ H. Reichel, Structural Induction on partial algebras, Akademie-Verlag, Mathematical research Vol.18, Berlin, 1984.

/REIT 78/ R. Reiter, On closed world databases, In: Logic and Databases (eds. H. Gallaire, J. Minker), Plenum Press, New York, 1978, 55-76.

/RISS 78/ J. Rissanen, Theory of joins for relational databases - a tutorial survey. LNCS 64, 1978, 537-551.

/ROKB 87/ M.A. Roth, H.F. Korth, D.S. Batory, SQL/NF: A query language for non1NF relational databases. Inform. Systems, 12, 1, 1987, p. 99-114.

/ROKS 85/ M.A. Roth, H.F. Korth, A. Silberschatz, Extended algebra and calculus for non-1NF relational databases. Revised Technical Report 84-36, Computer Science Department, University of Austin, 1985.

/SACC 85/ D. Sacca, Closures of Database Hypergraphs. Journal of ACM 32, 4, 1985, 774-803.

/SAUL 82/ A. Sadri, J.D. Ullman, Template dependencies: a large class of dependencies in relational databsaes and its complete axiomatization. Journal of ACM 29, 2, 1982, 363-372.

/SAWA 82/ Y. Sagiv, S. Walecka, Subset dependencies and a completeness result for a subclass of embedded multivalued dependencies. Journal of ACM, 29,1, 1982, 103-117.

/SCHS 84/ H.-J. Schek, M. Scholl, An algebra for the relational model with relation-valued attributes. Technical report DVSI-1984-T1, Technical University of Darmstadt, 1984.

/SCIO 81/ E. Sciore, Real-world MVD's. ACM SIGMOD Conference, 1981, 121-132.

/SCIO 82/ E. Sciore, A complete axiomatization for full join dependencies. Journal of ACM 29, 2, 1982, 373-393.

/SCOR'82/ E. Sciore, Inclusion dependencies and the universal instance. Technical report 82/041, SUNY at Stony Brook, Dept. of Comp. Sci.

/SDPF 81/ Y. Sagiv, C. Delobel, D.S. Parker, R. Fagin, An equivalence between relational database dependencies and a fragment of propositional logic. Journal of ACM 28, 3 (July 81), 435-453.

/SETH 85/ O. Selesnjew, B. Thalheim, On the number of minimal keys in relational databases over nonuniform domains. Acta Cybernetica, Szeged, 8, 3, 1988, 267-271.

/SHOK 86/ R.C. Shock, Computing the minimum cover of functional dependencies. Information Processing Letters 22, 3, 1986, 157-159.

/SMSM 77/ J.M. Smith, D.C.W. Smith, Data base abstractions: Aggregation and generalization. ACM TODS 2, 2, 1977.

/SOLO 78/ N.A. Solovjev, Testi, structura, teorija, primenenije. Nauka, Novosibirsk, 1978 (in Russian).

/SPER 28/ E. Sperner, Ein Satz Uber Untermengen einer endlichen Menge. Mathematische Zeitschrift 27 (1928), 544-548.

/SPYR 82/ N. Spyratos, A homomorphism theorem for data base mappings. Inf. Proc. Letters, 15, 11, Oct. 82, 91-96.

/STET 71/ S.J. Stephen, Y.S. Tang, An efficient algorithm for generating complete test sets for combinatorial logic circuits. IEEE Trans. Comput., 1971, C-20, 11, 1245 -1251.

/STPA 84/ A.A. Stognij, W.W. Pasitschnik, Reljazionnije modeli bas dannich. Institut Kibernetiki, Kiew 1984 (in Russian).

/SUMI 87/ Subieta K., M. Missala, Semantics for the entity-relationship model. The Entity-Relationship Approach,ed. by S. Spaccapietra, North-Holland, Amsterdam, 1987, 197 - 216.

/TAKY 79/ Y. Tanaka, Y. Kambayashi, S. Yajima, Properties of embedded multivalued dependencies in relational data bases. Trans. IEEE Japan E 62, 8, Aug. 1979, 536-543.

/THAL 83/ B. Thalheim, Decompositions in relational databases Colloquia Mathematica Societatis Janos Bolyai 42; Algebra, Combinatorics and Logic in Computer Science, Gyor, Hungary, 1983, 811-821.

/THAL 84/ B.Thalheim, Abhängigkeiten in Relationen. Dissertation (B), Technische Universität Dresden, 1985.

/THAL'84/ B. Thalheim, Deductive basis of relations. Proc. MFSSSS 84, LNCS 215, p. 226-230.

/THAL"84/ B. Thalheim, A complete axiomatization of full join dependencies. Bull. EATCS 24, 1984, p. 109-116.

/THAL 85/ B. Thalheim, Funktionale Abhängigkeiten in relationalen Datenstrukturen. J. Inf. Process. Cybern. EIK, 21, 1/2, 1985, p. 23-33.

/THAL 86/ B. Thalheim, Decomposition in relational databases. Proc. Coll. Algebra, Combinatorics and Logic in Computer Science , Colloqia Mathematica Soc. J. Bolyai, V. 42, North-Holland, 1985, p. 811-821.

/THAL'86/ B. Thalheim, A review of research on dependency theory in relational databases. Proc. 9th Int. Sem. on Database Management Systems, 1986, p. 136-159.

/THAL" 86/ B. Thalheim, Bibliographie zur Theorie der Abhangigkeiten in relationalen Datenbanken, 1970-1984, TU Dresden 566/85, Dresden 1985.

/THAL 87/ B. Thalheim, Design tools for large relational database systems. Proc. MFDBS-87-Conf., LNCS 305, p. 210-224.

/THAL~ 87/ B. Thalheim, Many-sorted variables in many-sorted logics. Submitted for publication.

/THAL'87/ B. Thalheim, On the number of keys in relational databases. Proc. FCT-87-Conf., Kazan, LNCS 1987.

/THAL"87/ B. Thalheim, Moderne Aspekte der Theorie der relationalen Datenbanken. X. Nullwerte in relationalen Datenbanken - eine Übersicht. Unpublished manuscript 1987.

/THAL 88/ B. Thalheim, Research on theory of generalized relational data bases. Unpublished manuscript, Kuwait University, Dept. of Mathematics, June 1988.

/THAL'88/ B. Thalheim, A systematic approach to database theory. Proc. INFO-88, 1988, p.

/THAL"88/ B. Thalheim, On semantic issues connected with keys in relational databases permitting null values. Journal Inf. Processing and Cyb., 24, 1988.

/THAL 89/ B. Thalheim, Logical Relational Database Design Tools Using Different Classes of Dependencies. Journal for New Generation Computer Systems, 1988, 1, 3, 1-18.

/THYA 88/ B. Thalheim, M. Yaseen, Data Base Modelling and Data Base Management Systems. Book submitted for publication, Kuwait 1988.

/TRA 50/ B.A. Trachtenbrot, Impossibility of an algorithm for the decision problem on finite classes, Dokladi akademii nauk 70, 1950, 569-572.

/TSLO 82/ D.C.Tsichritzis, F.H. Lochovsky, Data models. Prentice-Hall 1982.

/ULLM 80/ J.D. Ullman, Principles of database systems, Computer Science Press, Rockville, 1980.

/VARD 81/ M.Y. Vardi, The decision problem for database dependencies. Information Processing Letters 12,5, 1981, 251-254.

/VARD 84/ W.Y. Vardi, The implication and finite implication problems for typed template dependencies, Journal of Computer and System Sciences, 28,1, 1984, 3-28.

/VASH 78/ V.P. Vashenko, Multiple separation of a function using a fixed adjoint function. Soviet Math. Dokl. Vol.19 (1978), No.2, 246-249.

/VASS 80/ Y. Vassiliou, Functional depedencies and incomplete information. Proce. 6th Int. Conf. VLDB, 1980, 260-269.

/VIAN 83/ V. Vianu, Dynamic constraints and database evolution. 2nd ACM SIGACT-SIGMOD Symp. on Principles of Database Systems 1983, 389-399.

/VOIS 58/ J.K. Voischvillo, Metod uproschenija form vyrashenija funkzii istinosti. Naushnije dokladi vysschej schkoli, Filosofskije nauki, 1958, 2, 120 -135 (in Russian).

/VOSS 87/ G. Vossen, Datenbankmodelle, Datenbanksprachen und Datenbank-Management-Systeme. Addison-Wesley, Bonn, 1987.

/VTHI 84/ Vu Duc Thi, Remarks on closure operations. Kőzlemenyek 30, 1984, 73-87.

/YAPA 82/ M. Yannakakis, C.H. Papadimitriou, Algebraic depedencies. Journal of Computer and System Sciences 25, 1, Aug.82, 2-41.

/ZANI 76/ C. Zaniolo, Analysis and design of relational schemata for database systems. Technical report ULCA-ENG-7669, Los Angeles, 1976.

INDEX

aggregation	3
algebra	25
relational	25
antikey	105
anti-projection	27
Armstrong class	114
Armstrong database	114
Armstrong relation	114
attribute	7
characterization	
D	100
D2	128
F	91
strong D2	128
strong F	91
classification	3
closed class	128
closure	90
codependency	125
complement	26
full	30
restricted	26
conclusion	37
confluent	161
consequence	37
consistent	18
constraint	8
disjunctive existence	170
excluded functional	93
excluded multivalued	138
generalized functional	73
into	159
onto	159
union	194
cover	82
cross	59
full	125
data scheme	17
database	8
Armstrong	114
incomplete	169
with null values	169
database model	3
entity-relationship	20
hierarchical	24
network	24
object-oriented	31
Pawlak	31
relational	8

database scheme 8
- equivalent 15
- included 15
- relational 7
- weakly equivalent 15
- weakly included 15

DD 44
DEC 170
deductive basis 64
dependency 42
- afunctional 94
- algebraic 33
- antifunctional 93

- bounded domain 62
- Boolean 73
  - positive 73
- BV- 44

- closure 47
- co- 125
- conditional functional 190
- conditional inclusion 193
- conditional multivalued 193
- contextual join 125
- cross 59

- decomposition 44
- degenerated multivalued 121
  - full 121
- domain 62

- embedded 55
- embedded implicational 44
- embedded join 48
- embedded multivalued 45
- embedded projected join 48
- embedded template 45
- embedded tuple generating 45
- equality generating 45
- exclusion 158
- extented transitive 48

- first-order hierarchical 59
  - full 125
- functional 45
  - dual 75
  - monotone 75
  - positive 73
  - properties of 86
  - strong 75
  - weak 75

- general domain 164

dependency
- general embedded implicational 44
- general functional 45
  - normalized 61
  - properties of 60
- generalized closure 47
- generalized functional 45
  - properties of 73
- generalized multivalued 125
- generalized mutual 125
- graphical 125
- group 199

- Hungarian 97

- imposed functional 192
- inclusion 44
  - full 164
  - nondeterminstical 158
  - unary 154
  - weak 154
    - acyclic 157
    - typed 157

- join 34
  - acyclic 142
  - binary 125
  - m-ary 48
  - minimal 200
  - stronger 127
  - supercyclic 200

- key 76

- many-sorted 44
- mixed 125
- monadic 199
- multivalued 45
- mutual 125

- numerical 62

- predicative 51
- projected join 48
- propositional 82

- root 126

- subset 58

- template 44
  - bounded 65
  - properties of 51
- total BV- 44
- total projected join 48

dependency
    transitve 48
    tree 125
    tuple-generating 44
        properties of 51

    unrestricted functional 193
    uni-relational 44
        input stable 69
        input unstable 69

    weak multivalued 138

design by example 68
diagram 14
difference 27
dispersion 96
domain 7
    function 7
    independent 42

effluent 161
EGD 45
EMVD 45
entity 18
    scheme 18
    set 18
entity-relationship
    database 20
    diagram 20
    model 22
    scheme 16
ETGD 44
ETD 44
expression
    i 32
    relational 31
extension 28

FD 45
formal system 37
    complete 37
    full 39
    Hilbert-type 37
    sound 37
formula 10
    bounded 65
    definite 43
    equality 45
    follows 12
    generalized equality 45
    satisfy 11
    valid 12

functional dependency
    implication 193
    set implication 193

GD 45
generalization 3
GEID 44
GFD 45
GID 44
goal 189

horizontal decomposition 189

ID 49
implication 36
    finite 36
    k-ary 36
IND 44
inference rules 37
integrity constraint 8
interpretation 11
intersection 26

JD 48
join
    natural 28
    theta 28

key 83
    foreign 159
    minimal 102
    possible 174
    shortest 109
    sure 174
key-invariant 161
key set 181
    maximal 185
    reduced 182

limit 65

model 12
MVD 45

normal form
    deductive 65
    goal 189
null value 167
    context-dependent 179

projection 26
quotient 28
PJD 48
product
    Cartesian 28
proposition 82

relation 8
    Armstrong 114
    extended 164
    scheme 7
        nontrivial 56
        compatible 8
        non-uniform 109
        uniform 109
relationship 18
    complexity 20
    ID 22
    IS-A 22
    one-to-one 21
    one-to-many 21
    many-to-many 21
restriction 27
root 80

scheme 20
    recursive 21
selection 27
semantic system 37
    axiomatizable 37
        finitely 38
        k-ary 38
        U-bounded 38
semantic theory 37
Sheffer set 66
system
    closed 81
    independent 80
    Sperner 104

TD 44
TGD 44
tuple 8
    extended 169

union 26
union-invariant 161

variable
    free 10

Diese Reihe wurde geschaffen, um eine schnellere Veröffentlichung mathematischer Forschungsergebnisse und eine weitere Verbreitung von mathematischen Spezialvorlesungen zu erreichen. TEUBNER-TEXTE erscheinen in deutscher oder in englischer Sprache. Um Aktualität der Reihe zu erhalten, werden die TEUBNER-TEXTE im Manuskriptdruck hergestellt, da so die geringeren drucktechnischen Ansprüche eine raschere Herstellung ermöglichen. Autoren von TEUBNER-TEXTEN liefern an den Verlag ein reproduktionsfähiges Manuskript. Nähere Auskünfte darüber erhalten die Autoren vom Verlag.

This series has been initiated with a view to quicker publication of the results of mathematical research-work and a widespread circulation of special lectures on mathematics. TEUBNER-TEXTE will be published in German or English. In order to keep this series constantly up to date and to assure a quick distribution, the copies of these texts are produced by a photographic process (smalloffset printing) because its technical simplicity is ideally suited to this type of publication. Authors supply the publishers with a manuscript ready for reproduction in accordance with the latter's instructions.

Cette série de textes a été créée pour obtenir une publication plus rapide de résultats de recherches mathématiques et de conférences sur des problèmes mathématiques spéciaux. Les TEUBNER-TEXTE seront publiés en langues allemande ou anglaise. L'actualité des TEUBNER-TEXTE est assurée par un procédé photographique (impression offset). Les auteurs des TEUBNER-TEXTE sont priés de fournir à notre maison d'édition un manuscrit prêt à être reproduit. Des renseignements plus précis sur la forme du manuscrit leur sont donnés par notre maison.

Эта серия была создана для обеспечения более быстрого публикования результатов математических исследований и более широкого распространения математических лекций на специальные темы. Издания серии ТОЙБНЕР-ТЕКСТЕ будут публиковаться на немецком или английском языках. Для обеспечения актуальности серии, ее издания будут изготовляться фотомеханическом способом. Таким образом, более скромные требования к полиграфическому оформлению обеспечат более быстрое появление в свет. Авторы изданий серии ТОЙБНЕР-ТЕКСТЕ будут предоставлять издательству рукописи, удовлетворящие требованиям фотомеханического печатания. Более подробные сведения авторы получат от издательства.

B. G. Teubner Verlagsgesellschaft Stuttgart · Leipzig
O-7010 Leipzig, Postfach 930